8th International Scientific Conference on Building Defects (Building Defects 2016)

MATEC Web of Conferences Volume 93 (2017)

Ceske Budejovice, Czech Republic
24 - 25 November 2016

Editors:

Ingrid Juhasova Senitkova

ISBN: 978-1-5108-3510-8

Printed from e-media with permission by:

Curran Associates, Inc.
57 Morehouse Lane
Red Hook, NY 12571

Some format issues inherent in the e-media version may also appear in this print version.

Printed by Curran Associates, Inc. (2017)

For additional information, please contact EDP Sciences – Web of Conferences at the address below.

EDP Sciences – Web of Conferences
17, Avenue du Hoggar
Parc d'Activité de Courtabœuf
BP 112
F-91944 Les Ulis Cedex A
France

Phone: +33 (0) 1 69 18 75 75
Fax: +33 (0) 1 69 28 84 91

contact-edps@webofconferences.org

Additional copies of this publication are available from:

Curran Associates, Inc.
57 Morehouse Lane
Red Hook, NY 12571 USA
Phone: 845-758-0400
Fax: 845-758-2633
Email: curran@proceedings.com
Web: www.proceedings.com

TABLE OF CONTENTS

Aims & scope

MATEC Web of Conferences is an open-access journal devoted to the publishing of conference proceedings dealing with Materials science, Engineering and Chemistry.

The journal aims to make conference proceedings visible, citable and widely usable by the scientific community. It offers outstanding services for the publication and dissemination of contents in order to ensure long-term archiving and maximum exposure of scientific events.

Coverage includes all engineering disciplines: civil, naval, mechanical, chemical, and electrical engineering as well as nanotechnology and metrology. The journal also deals with the physical-chemical characterization of materials, their implementation, their resistance in their environment, etc. Others subdisciplines of chemistry, such as analytical chemistry, petrochemistry, organic chemistry, etc., and even pharmacology, are also welcome.

Why publish with MATEC Web of Conferences?

- **Open access:** conference papers available worldwide.
- **Fast publication:** less than 3 months.
- **Total flexibility:** no page restrictions, free unlimited use of colour in the online version, any kind of scientific document welcomed (traditional articles, posters, abstracts and slideshows).
- **High visibility:** all documents indexed in main databases and easily found online.
- **Fair author rights:** CC-BY Creative Commons license.
- **Range of additional publishing options:** printed proceedings, CD-ROMs, copy-editing, etc.

Contact: Isaline Augusto
matec@webofconferences.org

ISSN (electronic edition): 2261-236X

www.matec-conferences.org

edp open Open Access Publishing from EDP Sciences

International Scientific Committee

Organizing Committee

- **Terezie VONDRÁČKOVÁ, MSc., Ph.D.**
  ITB in České Budějovice, Czech Republic
- **Jan PLACHÝ, MSc., Ph.D.**
  ITB in České Budějovice, Czech Republic
- **Michal KRAUS, MSc., Ph.D.**
  ITB in České Budějovice, Czech Republic
- **Vladimír NÝVLT, MSc., MBA, Ph.D.**
  ITB in České Budějovice, Czech Republic
- **Josef MUSÍLEK, MSc., Ph.D.**
  ITB in České Budějovice, Czech Republic
- **Zuzana KRAMÁŘOVÁ, MSc., Ph.D.**
  ITB in České Budějovice, Czech Republic

Sponsors

- **Czech Chamber of Certified Engineers and Technicians (CKAIT)**
  Conference general partner
- **Czech Institution of Structural and Civil Engineers (CSSI)**
  Conference general partner
- **Journal Střechy-Fasády-Izolace**
  General medial partner
- **Internet portal Izolace.cz**
  Medial partner
- **Internet portal iMaterialy.cz**
  Medial partner
- **Journal Build Info**
  Medial partner

Photo Gallery

V.I.P.

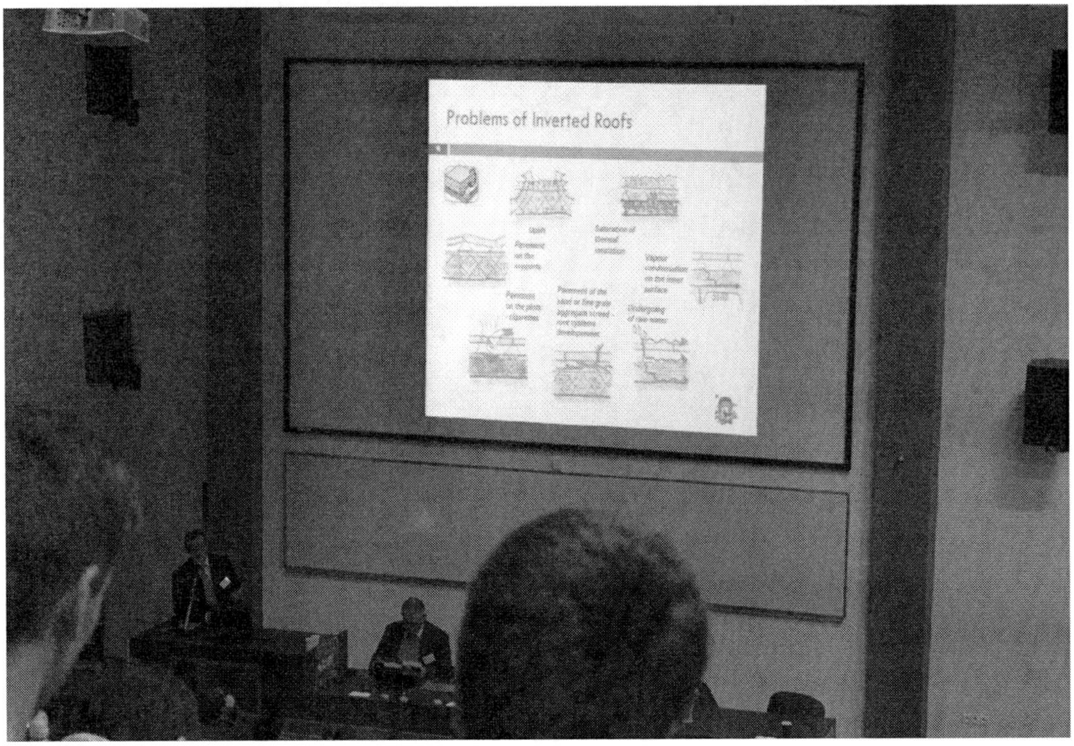
Problems of Inverted Roofs

NÁS

Statement of Peer review

In submitting conference proceedings to MATEC Web of Conferences, I certify to the Publisher that I adhere to the **Policy on Publishing Integrity** of the journal in order to safeguard good scientific practice in publishing.

1. All articles have been subjected to peer review administered by the proceedings editors.
2. Reviews have been conducted by expert referees, who have been requested to provide unbiased and constructive comments aimed, whenever possible, at improving the work.
3. Proceedings editors have taken all reasonable steps to ensure the quality of the materials they publish and their decision to accept or reject a paper for publication has been based only on the merits of the work and the relevance to the journal.

Title, date and place of the conference

8th International Conference Building Defects (Building Defects 2016)
24th – 25th November 2016

The Institute of Technology and Business in České Budějovice
České Budějovice, Czech Republic

Proceedings editor(s):

prof. Ingrid JUHÁSOVÁ ŠENITKOVÁ, MSc., CSc.

Date and editor's signature 25/11/2016

Preface

We are pleased to welcome you to the conference proceedings of the 8th International Conference on Building Defects (Building Defects 2016) held on 24th – 25th November 2016 at the Institute of Technology and Business (ITB) in České Budějovice, Czech Republic. Conference is organized on behalf of His Magnificence prof. Marek Vochozka, MSc., MBA., Ph.D., rector of the Institute of Technology and Business in České Budějovice and under the honorary patronage of the Czech Chamber of Certified Engineers and Technicians (CKAIT) and the Czech Institution of Structural and Civil Engineers (CSSI).

The current conference is the eighth continuation of Building Defects conferences, which solves especially the causes of defects during building and afterwards defects of building and building construction. The conference is the premiere chance for the presentation of new advances and research results in the fields of civil engineering. The main aim of the conference is the presentation of scientific research results and its practical solutions in the field of civil engineering and architecture, taking into account the principles of sustainable development. Contemporary civil engineering issues and its future development. The comprehensive content of the conference has attracted immense attention and the wealth of information spread out over all papers could be useful to professionals working in the related fields. It is my pleasure to announce the participation of expert speakers from various universities and research institutes in this two-day event.

A lot of thanks are given to all authors of submitted papers for their participation. They contributed a great deal of effort and creativity to produce this work. We are happy they choose the 8th International Conference on Building Defects to present to their excellent work. Thanks also to all editors and reviewers, who donated substantial time from their busy schedules to carefully read and evaluate the submitted scientific papers.

The conference proceedings contains 28 papers which are divided into 3 sections: Mechanical and Materials Engineering, Building materials, and Design and Technology. Each contributed paper was rigorously reviewed by 2 anonymous reviewers who were drawn from the International Scientific Committee.

At last but not at least, we would like to appreciate to all supporting organizations for their support and encouragement and got making the event a success. In addition, I would like to thank also to Organizing Committee for their great preparing work.

We are looking forward to the ninth International Conference Building Defects 2017 that will be held next year on November at the same location. We hope that it will be interesting and enjoying at least as all of its eight predecessors.

prof. Ingrid Juhásová Šenitková, MSc., CSc.
Conference Guarantee
ITB in České Budějovice

Fatigue failure mechanisms of thin-walled hybrid plate girders

Pavol Juhás[1,*]

[1]Institute of Technology and Business in České Budějovice, Department of Civil Engineering, Okružní 517/10, 37001 České Budějovice, Czech Republic

Abstract. Steel plate girders of bridges and industrial structures are usually subjected to dynamic fatigue loading. Their thin-slender webs under dynamic fatigue loading are laterally buckling and vibrating. The consequence of that can be arising and successive development of fatigue cracks along the welded connections of the thin webs with flanges and stiffeners. Therefore, the fatigue strength and life/time of thin-walled plate girders depend beside of usual material-technological and structural influences also on stability aspects. Basic knowledge about fatigue failure mechanisms of thin-walled hybrid plate girders is presented with the regard for their local stability, obtained and by theoretical an experimental analyses. The outcome of obtained results is that the fatigue consequences of thin webs vibration can be effectively reduced by appropriate reduction of their slenderness. Therefore, the empirical formula for the limitation of web slenderness has been proposed and recommended for plate girders subjected to dynamic fatigue loading.

1 Introduction

Generally, it is well known, that behaviour of plate girders depends in large degree on local stability of their flanges and webs. New standards (EN 1993-1-1) classify the cross-sections of plate girders like compact (class 1), quasi compact (classes 2 and 3) or slender (class 4) ones relating to local stability. In the case of compact and quasi compact cross-sections the flanges and webs are stable or relative stable, the stress-strain relations are linear and the simplified calculation procedures can be used. In the case of slender cross-sections, the flanges, webs, or flanges and webs are buckling in dependence on geometrical and material properties and on the level and mode of loading. Therefore, the stress-strain relations are non-linear and the load-carrying of plate girders is decreasing. But the interaction of webs, flanges and stiffeners in individual fields of plate girders has beneficial effect on their load-carrying capacity load-carrying capacity. Consequently, the actual load-carrying capacity can be generally much higher than theoretical critical capacity – buckling strength. This reflects the post critical behaviour and calculation procedures on the stiffened plate girders with slender cross-sections [1-4].

* Corresponding author: 15364@mail.vstecb.cz

Post critical calculation procedures are based on the research of plate girders subjected mainly to static or quasi-static loading. But plate girders are used in many cases for bridges and industrial structures that are subjected to dynamic fatigue loading. This phenomenon reduced the fatigue strength and the life-time of plate girders with slender cross-sections. Therefore, it is very important to know the real load-carrying capacity and behaviour of such plate girders during actual loading process.

The design of plate girders with slender cross-sections, particularly with slender webs is convenient from economical point of view. Hybrid plate girders with slender web and higher strength steel in flanges are especially advantageous in this direction. The paper deals with fatigue failure mechanisms of thin-walled hybrid plate girders from local stability aspects [4-9].

2 Experimental investigations

The complete experimental programmes consisted from static and dynamic fatigue tests of 48 plate girders with symmetrical I cross-section of real practical dimensions produced under usual technological conditions. The tested girders were different in geometrical dimensions, stiffened field ratioα, web slenderness β and combination of flanges and web steels (material and cross-section groups). One half of the tested girders were used for static tests and the other half of similar girders for dynamic fatigue tests [6, 8-9].

The cross-section dimensions in the individual groups were selected to give (at uniform t_w thickness and various b_w width) slenderness ratio β_w = 120, 150 and 180. The flange dimensions were designed to ensure local stability. Element lengths were linked to b_w web width ($L \approx 7b_w + 800$ mm). The intermediate panels had different side ratios ($1 \leq \alpha \leq 3.5$).

Steels of flanges were: 11 523.1, 13 221.1, 15 422.5 and 16 224.1, steel of webs was uniform – 11 375.1 (by previous Czech and Slovak material standards).

Scheme of girders and working arrangement of the tests are shown in Fig. 1.

Fig. 1. Scheme of girders and working arrangement of the tests.

The flanges and webs were machine flame cut. Besides of that the tension flanges of girders were planned. The fillet welds between flanges and webs were made by automatic machine under flux, while stiffeners were hand welded. Prior of testing the true geometrical and material characteristics of the flanges and webs were ascertained. All steels used for the experiments were in conformance with the requirements of applicable standards, or exceeded them. The geometrical and material characteristics, as well as their corresponding deviations from design dimensions and standard material specifications are random variables. Hence, in evaluations of experimental results, either the directly determined or those generally probabilistically - statistically defined values of these variables could be used.

The tested girders were sustained by special vertical and horizontal supports to prevent lateral buckling. The loading was realised by two hydraulic jacks. The tests of girders were divided into two parts. First of all, the girder was loaded by successive loading till to P_{max} and after that about five cycles between $P = 0$ and $P = P_{max}$ were repeated. Then the fatigue test could start. During fatigue test the girders were loaded by synchronise pulsate loads

$$P(t) = P_0 + P_1 \sin(2\pi f t), \quad (1)$$

$$P_0 = 0.5\,(P_{max} + P_{min}),\ P_1 = 0.5\,(P_{max} - P_{min}), \quad (2)$$

$$P_{min} = (0.2 \div 0.4)P_u\,,\ P_{max} = (0.7 \div 0.8)P_u\,, \quad (3)$$

P_u is the ultimate static loading including the elastic-plastic post critical behaviour of the web, frequency $f = 5.0$ Hz.

The behaviour of girders during the static and dynamic tests was ascertained by measurement, registration and evaluation of strains ε in the most stressed parts (78 strain gauges). The supports sets down v_0 - deflection pickups 1 and 5, deflection of girder v_s in the middle cross-section - deflection pickup 3 and buckling of the web w in the middle points of two stiffened panels - deflection pickups 2 and 4 were also measured. Besides of that we made the detailed measurement of static and dynamic buckling of the web w in different points and cross-sections (I, II, ..., XXII) by movable pickup with graphic evaluation, or by inductive pickup, strain gauge bridge and oscillograph, respectively.

The fatigue tests were realised in stages. After each 50000 cycles the tests were interrupted. Incidental arising of the cracks was ascertained by the detailed observation of the girder with the help of magnifying glass and by direct graphic and tabulate evaluation of strains ε and deflections w and v_s at loads $P = 0$, P_{min}, P_{max}. The differences between last and previous values of strains ε in some points of girder indicate arising and placement of fatigue cracks. After arising of the cracks, we followed their development and influence on the global behaviour of girders. The tests continued until the total failure occurred.

3 Behaviour of tested girders and basic knowledge

3.1 Formation of the shapes of the web buckling and vibration

Under quasi-static loading the thin webs of girders are laterally buckling in dependence on the initial imperfections, geometrical and material parameters and stiffeners of bordered flanges and stiffeners. They are buckling in shapes which correspond to the mechanism of their behaviour according to the process of loading. These shapes are formed already at the beginning of loading process. In accordance with the initial shape a new service shape can be formed by the successive fluent transition or the sudden jump over the web. The service shapes of buckling are decisive for analysis of load-carrying capacity and behaviour of the thin webs of plate girders. Under the dynamic fatigue loading the thin webs of girders are laterally vibrating. At the beginning of loading process, the service shape of the web vibration is formed in a similar way to the case of quasi-static loading. During the usual service loading regime in girders the resonant effects do not arise. Therefore, the lateral web deflections which are received at the repeating loading are practically identical with the dynamic lateral web deflections. Under the fixed cyclic loading the levels of deformations, deflections and shape of lateral web vibration did not change at any tested girder till the beginning of fatigue failure. This knowledge is evident from Fig. 2.

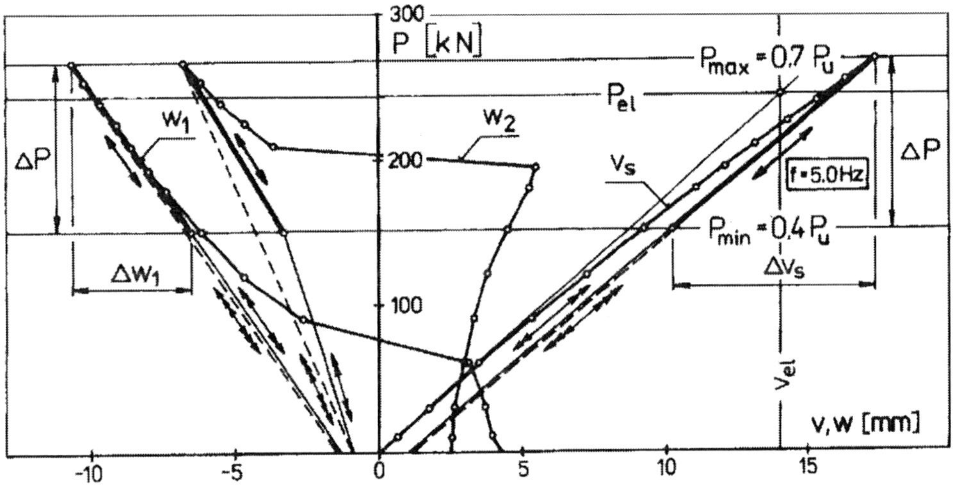

Fig. 2. Load – deflection relationships of the thin web and girder. Jumps over of the web in the most loaded fields.

Fig. 2 expresses the load-deflection relationships of the chosen girder w_s in the middle cross-section and its web w_1 and w_2 in the central points of the two most loaded fields. The ultimate load of the girder with account of the web buckling and the elasto-plastic postcritical behaviour of the web is P_u=382.2 kN. During the first loading the sudden jump over appeared in two most loaded web fields - in the web field S_{2-1} after P = 60 kN and in the web field S_{2-2} after P = 195 kN. However, during the following repeating loading between P = 0 and $P = P_{max}$ (quasi-static loading) or between $P = P_{min}$ and $P = P_{max}$ (dynamic fatigue loading) any next jump over of the web did not appeared. Under repeating loading relatively quick stabilisation of deformations and deflections was set in and the web buckled or vibrated in the service shape.

3.2 Formation and successive increase of the fatigue cracks

The results of realised experimental investigation also proved that the lateral web vibration of plate girders can also cause the formation and successively increase of the fatigue cracks which are situated near the welded connections of the web with flanges and stiffeners. Therefore, the fatigue strength and life-time of plate girders with thin web depend besides the usual loading, material and construction influences also on the lateral web vibration. Fatigue cracks of the chosen tested girder are presented in Fig. 3.

This girder is especially interesting with fatigue cracks along compressed flange in all internal web fields. The first cracks 1 and 2 had appeared after 300 000 and 400 000 cycles, respectively. Later cracks were successively increasing, but only the initiation of the crack 5 along stiffener was decisive for the failure of the girder. The crack had spread very

 DOI: 10.1051/matecconf/20179301001

quickly to the tension flange in the form of crack 6 and the total failure of girder was completed after 1 342 000 cycles.

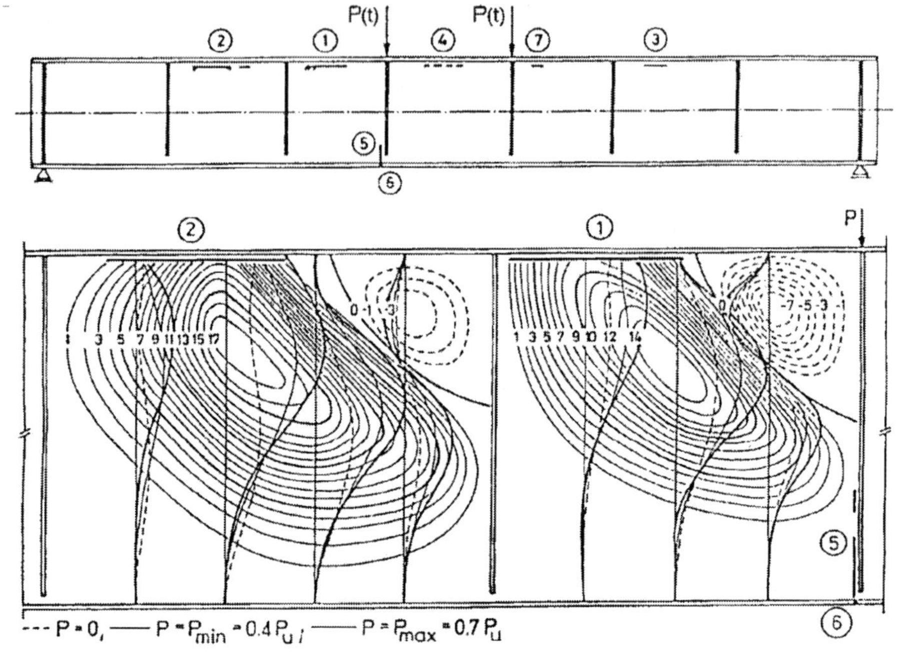

Fig. 3. Fatigue cracks in the thin web along welded connections with compression flange and transverse stiffener; crack in tension flange. Final topography of the most failed fields.

Three types of the fatigue cracks appeared in the tested girders. The crack of the type 1 (1, 2, 3, 4, 7) is in the web under the compression flange, it is spreading along the fillet welding. It can pass also into the web depending from the membrane stresses. The crack of the type 2 (5) is in the tension web part, it is passing along the stiffener to tension flange. The crack of the type 3 (6) is in the tension flange, it appears after developing of the crack of type 2. The intensity of the crack's propagation for different crack types is the important factor of the complex life-time evaluation of plate girders. Due to our results there is difference in the crack's propagation velocity. The slowest cracks were those of the type 1, the fastest were cracks of the type 3 and cracks that are the results of production defects.

3.3 Limit criterion for the design of plate girders subjected to fatigue loading

The realised investigation was especially directed on the fatigue cracks of type 1 and 2. The fatigue crack of type 1 is illustrated in Fig. 4. These cracks depend first of all on the secondary transverse bending stresses along flanges and on their changes during loading process. These bending stresses of the web can be reduced above all by their slenderness β. Therefore, the web slenderness should be adequately limited.

According to the results of realised investigation and taking into account the results of previous researches [7-10] the maximum slenderness of the web for the plate girders subjected mostly to bending is:

$$(280/\sqrt[3]{N})\sqrt{235/f_{yf}}\ , \tag{4}$$

N is needed, or required number of cycles and f_{yf} is yield stress of flanges.

Fig. 4. Fatigue crack of type 1, local stability failure of the compression flange.

4 Conclusions

The realised investigations have proved that the service shapes of buckling and vibration are decisive for the analysis of behaviour of thin-walled plate girders. The lateral web vibration of plate girders can cause formation and successive increasing of failure cracks which are situated near the welded connections of the web with flanges and stiffeners. Therefore, the fatigue strength of plate girders with thin web depends besides the usual loading, material and construction influences also on the lateral web vibration. Deformation and increasing of fatigue cracks in thin webs of plate girders depend especially on the bending stresses along flanges and stiffeners and on their changes during loading process. The bending stresses of the web can be reduced above all by their slenderness β. Therefore, the web slenderness should be adequately limited according to presented proposal.

References

1. L.S. Beedle, et al., Stability of Metal Structures - A World View, Headquarters, Bethlehem (1991)
2. T.V. Galambos, et al., Guide of Stability Design Criteria for Metal Structures - Structural Stability Research Council, J. W., New York (1998)
3. P. Juhás, et al., Design of Steel Structures – Commentary to STN 73 1401, SUTN, Bratislava (2001)
4. Y. Maeda, I. Okura, I., *Influence of Initial Deflection of Plate Girder Webs on Fatigue Crack Initiation*, Eng. Struct., **5**, 1, 58-66 (1983)
5. P. Juhás, *Thin-Walled Combined Plate Girders*, Stav. Čas., **35**, 12 (1987), p. 871-907
6. P. Juhás, *Fatigue Failure of Thin-Walled Combined Plate Girders*, Stav. Čas., **38**, 11, 819-847 (1990)
7. I. Okura, B.I. Yen, J.W. Fischer, *Fatigue Strength of Thin-Walled Plate Girders*, Struct. Engin. Inter., **1**, 1, 39-44 (1993)

8. P. Juhás, *Fatigue Failure Mechanism of Steel Elements*, Extending of Lifespan of Structures, IABSE Symposium, San Francisco, 1085-1090 (1995)
9. P. Juhás, E. Juhásová, Local Stability and Load-Carrying Capacity of Plate Girders Subjected to Repeated Loading, Proc. of Struct. Stability Research Council – Annual Technical Session and Meeting, Memphis, 257-266 (1998)

Defects of large diameter drilled piles

Jaroslav Žák[1,*] and *Jiří* Žák [2]

[1]Department of Civil Engineering, Institute of Technology and Business in České Budějovice, Okružní 517/10,370 01 České Budějovice, Czech Republic
[2]Faculty of Science, Masaryk University in Brno, Kotlářská 267/2, 611 37 Brno, Czech Republic

Abstract. High-rise buildings are often based on large-diameter bored piles. Their flaws and faults then affect the behavior of the entire building and can significantly influence the statics. Correct processes is therefore necessary to pay attention. The article describes the most common wrong ways of implementing the recommendations and how to prevent them. The design of piles is based on calculated loads, respectively forces and moments acting on the piles and of parameters of the environment. There may, of course, committed a number of mistakes, but we further assume piles to be designed in accordance with all applicable standards. We will focus mainly on technological progress of the implementation of drilling and concreting piles and show the most common mistakes. Bored piles can be performed using a casing or without them. In most cases, the use of casings is possible. Recommendations how to ensure the quality and to protect the environment are also given.

1 Introduction

A designer must take into a consideration calculated strain that is what forces and moments are acting upon piles and also parameters of the surroundings (subgrade) when designing a pile. One may make several mistakes here however let's assume properly designed pile which meets all the required standards. We will not discuss the actual propagation of the calculated bending moment from the base of the pile a whether moments calculated along the length of the pile are correct. Let´s also assume that the longitudinal and spiral reinforcement was correctly designed.

Further we will discuss mostly technological procedure of drilling and concreting piles and we will point out the most common mistakes. Bored piles can be performed using a casing or without them. In most cases, the use of casings is possible. The author recommends their usage due to ensuring of higher quality of concrete and also as environment protection.

2 Theoretical formulas

There are 3 main equations which can be used to describe and simulate flow of concrete paste, see [1-5] for more details. The main equations which is being used is the equation of

* Corresponding author: jzax@seznam.cz

motion. Others are the continuity equation and the energy equation. According to the type of approximation and whether we consider concrete paste to be a Newtonian fluid or non-Newtonian fluid, Navier-Stokes equation or the Cauchy equation of motion can be used. Cauchy equation of motion is listed below (1), (2)

$$\rho\left(\frac{\partial v}{\partial t}+v\cdot\nabla v\right)=\nabla\cdot\sigma+\rho g \tag{1}$$

The term **v** is the 4 dimensional (time being one dimension) and 3 directional velocity, ρ is the density, t is the time, **g** is the gravity and $\boldsymbol{\sigma}$ is the (total) stress tensor. The stress tensor is given by

$$\sigma=-\rho I+T \tag{2}$$

In general and as shown above, the stress tensor is decomposed into an isotropic pressure term ρ and an extra stress contribution **T**, commonly known as the extra stress tensor. **I** is the unit dyadic.

Second equation is the continuity equation (3) which is often used to calculate pressure correction while solving Eq. (1). It writes

$$\frac{\partial v}{\partial t}=-(\nabla\cdot\rho v) \tag{3}$$

For liquids, which are much denser than gasses, the incompressibility constraint is often assumed, and the above equation can be written as

$$(\nabla\cdot v)=0 \tag{4}$$

The last of the governing equations is the energy equation, which is used to determine possible temperature variations. However, during flow of cement based materials, isothermal conditions are often assumed and the energy equation is often omitted.

3 Technological procedure

Mistakes can be found even in the technological regulations; however most common mistakes are seen because the regulation were not properly met. Distance wheels from various materials are being used when constructing piles. Wheels not properly attached to the main reinforcement. Designed diameter of piles and reinforcement does not allow smooth insertion into the drilled hole. Concreting is being carried out in a wrong way, finishing of the concreting is being carried out in the wrong way. Samples of the concrete are being taken from improper places (not risky locations) and are not being taken from the risky places.

3.1 Cover of a reinforcement

To ensure cover of reinforcement distance wheels can be used (see Fig. 1). Plastic distance wheels turned out to be unsuitable because of their deformation while placing. Insufficient attention is often paid to manufacturing wheels, they are treated as not important and auxiliary part of the construction. Their incorrect manufacture can however cause serious issues. Distance wheels are sometimes manufactured with inner diameter too big and its centre is not located within centre of the wheel. Afterwards distance wheel can crush, break or damage itself and can cause problems for smooth insertion of pile reinforcement.

Fig. 1. Distance wheel of a pile reinforcement.

Distance wheels with too big or unsymmetrical opening must be removed from the construction.

It is often the case that the distance wheels are not placed correctly on the pile reinforcement. Designers often design diameter of the wheel as a standard cover. Because of this the wheel must be placed on the outer edge of the spiral. Also a designer can design diameter of the drill expanded only by casing insertion of pile reinforcement is practically infeasible. Reaction of a contractor is often placing the distance wheels on the inner edge of the spiral (see Fig. 2) however by doing this the actual cover is reduced. You can clearly see that reinforcement is in the contact with the casing (see Fig. 2). Casing will not be working properly or even it can happen that after removal of the casing the reinforcement can be moved from the concrete. Early onset of corrosion is inevitable.

Pile reinforcement seems to be rigid but it has tendency to undulate along its length. Pile reinforcement should be designed to be sufficiently rigid due to better manipulation when inserting. Designers often design only three wheels in the cut and reinforcement of smaller profiles in order to cut costs. Mathematically it is sufficient to ensure proper cover but not realistically. Undulation of pile reinforcement, deformation and movement of wheels can cause that the pile reinforcement may be in touch with the casing. Wheels must be therefore placed in single cut and turned by 60 degrees to each other. Wheels must be placed on the outer side and the diamcter of the drill must be designed with sufficient reserve allowing good manipulation with pile reinforcement in drill hole. Each layers of the wheel must not be too far away and must be turned by 60 degrees.

Fig. 2. Placement of the distance wheels.

3.2 Concreting

Concreting is being performed with a hopper and its extensions to the desired length to ensure the concrete will not undergo defragmentation due to a falling a large distance. Great time costs can happen to a constructer when following this procedure. Several of temporally removal of the hopper and adjusting the extension to desired length and then returning the hopper back during concreting is very time consuming. Constructors often use only the upper part of the hopper which makes the concreting much faster. However, the result is poor quality and nonhomogeneous concrete.

A very delicate matter is concreting the head of the pile. Water is often leaking through the casing during the concreting. Dirt is falling into the drilled hole and everything is being pushed upwards by the concrete paste. Therefore, it is necessary to concrete head of the pile with extension connected to the concrete paste a thus pushing the water and dirt to the surface. Concreting should be done only to the specified height which exceeds the head of the pile and after hardening of the concrete paste removal of the poor quality concrete is necessary.

While following this procedure several problems can occur during finishing of the concreting. Concrete paste might not be able to push the dirt through inserted extension upwards. Pulling out the extension happens to be often the solution however dirt and water will be released into the area of the head of a pile. Thus poor quality concrete will be located at the most stressing part of the pile. Afterwards shutter is being put on and the rest of the concreting is being done without extension with high quality concrete thus making an impression of high quality concrete within the head of the pile although there is poor quality concrete just below it. See Fig. 4 and 5.

Concreting must solely be done continuously without interruptions. It is necessary to push dirt above the head of the pile and remove any leftover concrete.

Fig. 3. Pile reinforcement.

3.3 Concrete sampling

There is concrete sampling before the concreting according to the relevant standards. During the concreting sampling occurs when in doubt. Common mistake is to take samples from the head of the pile after concreting interruption and using new high quality concrete paste.

Fig. 4. Pile concreting.

 DOI: 10.1051/matecconf/20179301002

The significance of sampling is to take sample during nonstandard procedures such as prematurely pulled out extensions, water and dirt detection in the concrete paste.

Fig. 5. Heads of pile concreting.

4 Conclusions

The purpose of this article was not to describe in details all the problems connected with defects and fails of large diameter drilled piles. It was to point out some of them. Based on his experience the first author recommends following:

- Usage of concrete distance wheels, precisely geometrically manufactured and correctly placed on pile reinforcement.
- To have each layer of distance wheel rotated against each other.
- Thoroughly follow proper concreting procedure especially using sufficiently long extension.
- With regard to manipulation to design sufficient solid pile reinforcement.
- Removal of concrete above a head of the pile.
- Take samples in doubts.

References

1. F. Irgens, Continuum Mechanics (Springer Verlag, 2008)
2. G.E. Mase, *Schaums Outline Series: Theory and Problems of Continuum Mechanics*, (McGraw Hill Inc, 1970)
3. J.F. Wendt, *Computational Fluid Dynamics, An Introduction* (Springer, 1992)
4. H. Mori, Y. Tanigawa, *Simulation Methods for Fluidity of Fresh Concrete* (Mem. Of the School of Eng. Nagoya Un., 1992)
5. N. Roussel, *Steady and transient flow behaviour of fresh cement pastes*, Cem. Concr. Res. **35** (2005)

MATEC Web of Conferences **93**, 01003 (2017) DOI: 10.1051/matecconf/20179301003
Building Defects 2016

Reliability of timber structure bolt connection subjected to double unequal shears with thick plates as outer members

Josef Musílek[1,*] and *Jaroslav* Žák [1]

[1]Institute of Technology and Business in České Budějovice, Department of Civil Engineering, Okružní 517/10, 370 01 České Budějovice, Czech Republic

Abstract. For reliability check of timber structure bolt connections subjected to double shear with thick plates as outer members and inner timber member is nowadays possible to use equations which were derived by Johansen and which are also used in Eurocode standards. These equations assume that the loads are the same on both sides. But it may often be a case when the loads are not the same. The paper deals with reliability in ultimate limit state of timber structure bolted connections subjected to unequal double shear with thick plates as outer members and inner timber member. There are derived equations for calculation of the reliability of this connection type in the ultimate limit state.

1 Introduction

Structure engineers can use nowadays for reliability check of timber structure bolt connections subjected to double shear with thick plates as outer members and inner timber member Eurocode standards [1]. These formulas were derived by Johansen in [2] and can be found also for example I [3]

In [4] there were derived formulas for design load-carrying capacity of bolt connections subjected to double unequal shear with thick plates as outer members and inner timber member. Example of such connection is showed in Fig. 1. We assume that $F_{d1} > F_{d2}$. The formulas are listed below as Eq. 1, 2, and 3. Each formula corresponds to a particular mode of failure. In formula (1), the member for characteristic withdrawal capacity of the bolt is omitted.

The case when the thin plates as outer members are used can occur very often but this case is not the subject to this paper. This case is described in [5-7].

* Corresponding author: musilek@mail.vstecb.cz

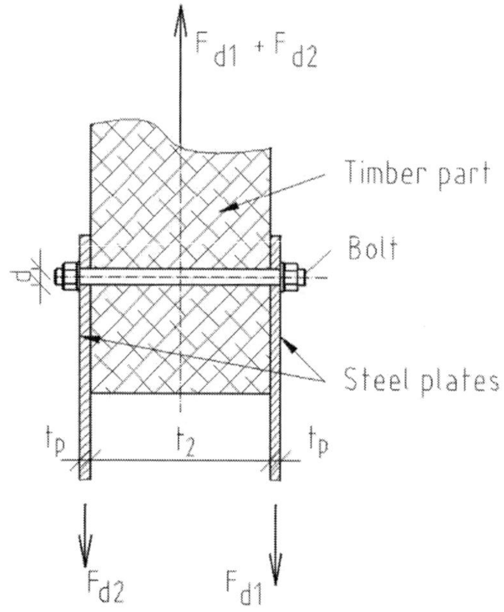

Fig. 1 Bolted connection subjected to double shear.

Mode of the failure a)

$$R_{d1} = \frac{k_{mod}}{\gamma_M}\left(2{,}3 \cdot \sqrt{M_{y,Rk} \cdot f_{h,2,k} \cdot d}\right) \tag{1}$$

Mode of the failure b)

$$R_{d1} = \frac{k_{mod}}{\gamma_M} \cdot f_{h,2,k} \cdot d \cdot t_2 - R_{d2} \tag{2}$$

Mode of the failure c)

$$R_{d1} = \frac{k_{mod}}{\gamma_M} \cdot f_{h,2,k} \cdot d \cdot \left(\sqrt{2 \cdot \left(t_2{}^2 + \frac{4 \cdot M_{yk}}{f_{h,2,k} \cdot d} + \frac{2 \cdot \gamma_M \cdot R_{d2} \cdot t_2}{k_{mod} \cdot f_{h,2,k} \cdot d}\right)} - t_2\right) - R_{d2} \tag{3}$$

2 Reliability of bolted connection subjected to unequal double shear according to Figure 1

We can assess the reliability in ultimate limit state according to Eq. 4

$$\frac{F_{d1}}{R_{d1}} \leq 1 \tag{4}$$

For determination of reliability is needed to define the way of loading. For mode of failure "a)" the design load carrying capacity R_{d1} is independent on the R_{d2} so the way of loading is unimportant. But for the modes of failure "b)" and "c)" the situation is different. Let's assume the connection according the Fig. 1 for: $f_{h,2,k}$ =*26,174MPa*, t_2 = *60mm*, d = *16mm*, k_{mod} = *0,8*, γ_M = *1,3*, $f_{u,k}$ = *360MPa*.

 DOI: 10.1051/matecconf/20179301003

In Fig. 2 the dependence R_{d1} on R_{d2} is shown for the mode of failure "b)" which is described by Eq.2.

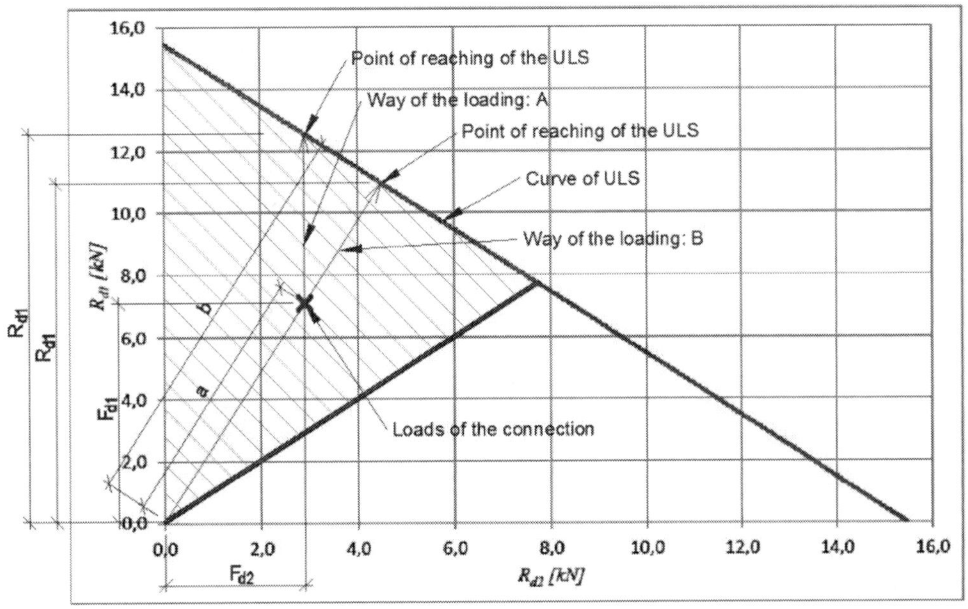

Fig. 2 Assessment of the connection reliability for mode of failure "b)" – Eq. 2.

In Fig. 3 the dependence R_{d1} on R_{d2} is shown for the mode of failure "c)" which is described by Eq.3.

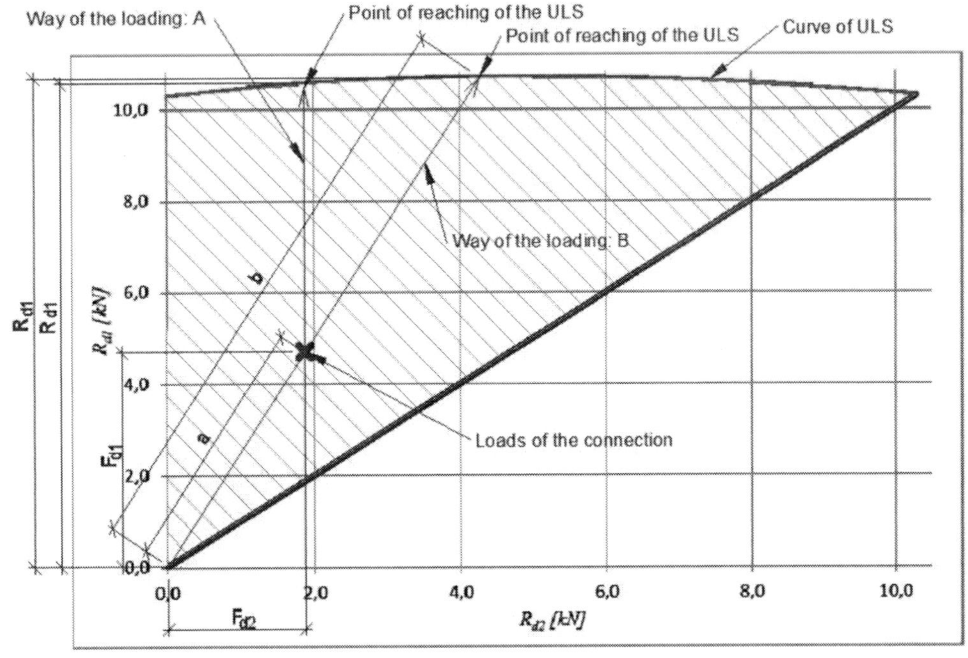

Fig. 3 Assessment of the connection reliability for mode of failure "c)" – Eq. 3.

The design loads of the connection F_{d1} and F_{d2} according to Fig. 1 can be drawn as a point in Fig. 2 and Fig. 3. Because we assume in accordance with 1 that $F_{d1} > F_{d2}$, the points representing the loads of the connection will be located always above the line with slope of 45°. If the point with coordinates $[X= F_{d2}\, ;\, Y= F_{d1}]$ is located under the red curve in hatch area, the connection is satisfied in terms of the ultimate limit state. So the red curve represents the reaching of the ultimate limit state. But for the assessment of the connection reliability is advisable to derive a formula, which would give us a percentage measure of the reliability connection. This percentage would tell us "how far" the connection design is to the ultimate limit state. This is the reason why we need to define the way of the loading as it was mentioned in the previous. In the following the two ways "A" and "B" of loading are considered.

3 Reliability for the way of the loading "A"

In this way of loading the design load F_{d2} stays constant and the design load F_{d1} is increasing only. Then we can define $R_{d2}= F_{d2}$ and substitute it into the Eq. 2 and 3 which we calculate the design load-carrying capacity R_{d1} from. The reliability condition of the connection in terms of the ultimate limit state is then:

$$\frac{F_{d1}}{R_{d1}} \leq 1 \tag{5}$$

The R_{d1} is taken as the minimum value got from Eq. 1,2 and 3

4 Reliability for the way of the loading "B"

In this way of loading, the both design loads are increasing and the ratio F_{d1}/F_{d2} remains constant. This way of loading is possible to expect in practice most often.

In accordance with Fig. 2 and Fig. 3, the assessment of the ultimate limit state can be expressed as:

$$\frac{a}{b} = \frac{F_{d1}}{R_{d1}} < 1 \tag{6}$$

Let´s define a ratio n:

$$n = \frac{F_{d2}}{F_{d1}} = \frac{R_{d2}}{R_{d1}} < 1 \quad => \; R_{d2} = n \cdot R_{d1} \tag{7}$$

When we substitute Eq.7 into Eq. 2 (which describes the failure mode "b)") and make some mathematical simplifications, we will get the design load-carrying capacity R_{d1} depending on n:

$$R_{d1} = \frac{k_{mod}}{\gamma_M} \cdot \frac{t_2 \cdot f_{h,2,k} \cdot d}{1+n} \tag{8}$$

And similarly, when we substitute Eq.7 into Eq. 3 (which describes the failure mode "c)"), we will get the design load-carrying capacity R_{d1} depending on n:

$$R_{d1} = \frac{k_{mod}}{\gamma_M} \cdot f_{h,2,k} \cdot d \cdot \frac{\sqrt{t_2{}^2 \cdot 2 \cdot (1+n^2) + (1+n)^2 \cdot \frac{8 \cdot M_{yk}}{f_{h,2,k} \cdot d}} + t_2 \cdot (n-1)}{(1+n)^2} \tag{9}$$

5 Summary

The formulas for assessment of reliability in terms of ultimate limit state of bolted connection subjected to unequal double shear with thick plates as outer members were derived.

References

1. *EN 1995-1-1, Design of timber structures. Part 1-1: General-Common rules and rules for buildings* (1995)
2. K.W. Johansen, Theory of timber connections. International Association of Bridge and Structural Engineering. **9,** 249-262 (1949)
3. J. Porteous, A. Kermani, *Structural timber design to Eurocode 5* (Blackwell Publishing, 2007)
4. K. Kubečka, J Musílek, *Contemporary Problems of Architecture and Construction* (Univerzita Degli Study Firenze, **7**, 59-62, 2015)
5. J. Musílek, K. Kubečka, AMM **752-753**, 711-71 (2015)
6. J. Musílek, K. Kubečka, AMM **752-753**, 652-655 (2015)
7. J. Musílek, K. Kubečka, AMM **752-753**, 648-651 (2015)

MATEC Web of Conferences 93, 01004 (2017) DOI: 10.1051/matecconf/20179301004
Building Defects 2016

Moisture diagnostic of a failured roof drainage situated behind a parapet wall and a subsequent condensate influence to a wooden construction

Josef Remeš[1,*], *Roman* Brzoň[1], *Tomáš* Petříček[1], and *Jiří* Horák[1]

[1]BUT, Faculty of Civil Engineering, Veveří 331/95, 602 00 Brno, Czech Republic

Abstract. There are several possibilities how to deal with a roof drainage. From an architectonic point of view an improper and unsuitable system is chosen in a majority of cases which may be as a logical consequence the beginning of upcoming problems. One of these problems is typically connected with either a roof location near the parapet wall, or in the corner of a flat roof. Another one is generally represented by a roof drainage solution with a side inflow for the drainage system that goes through the parapet wall construction. This solution is represented by a useful space increase in a layout of the building and an installation of a rainwater drainage core is not necessary in such cases. In case more of negative effects accumulate, such as an improper thermal insulation quality or insufficient thickness of the thermal insulation, significant problems may appear hand in hand with using of the building. These problems are generally represented by a moisture appearance in the inflow area. What more, this is not always detected and a major cause can be determined very hardly. This article deals with possible solutions for the wooden construction failure caused by the moisture that may emerge as an improper inflow layout consequence.

1 Introduction

Ambient humidity condensation may appear in almost any building type. If this kind of humidity appears only for a short period of time (e.g. during cooking time) there is no need to deal with this issue as it evaporates soon after. On contrary if there is either a long-term or extensive humidity a certain problem may appear. Humidity may bring degradation to the surface layers of plater or constructions. In this case ambient humidity condensation was related to a house where humidity maps started to appear due to an improper inflow layout and a combination of other influencing factors and as a subsequence adequate house living was not enabled [1-4].

* Corresponding author: remes.j@fce.vutbr.cz

DOI: 10.1051/matecconf/20179301004

1.1 Building description

The building is considered for living and occupied by five users. There is a 2-floor detached house situated in a north-east part of the municipality. The building is a composition of two mass in a rectangular shape. The first part consists of two-flow section roofed with a flat mono-pitched roof which is drained to a parapet wall. The parapet wall is subsequently led to the one-floor area roof. The one-floor part roof is a flat roof loaded by pea gravel, drained by two roof drains which are situated above a living room and bedroom of the house.

Fig. 1. Elevation at building.

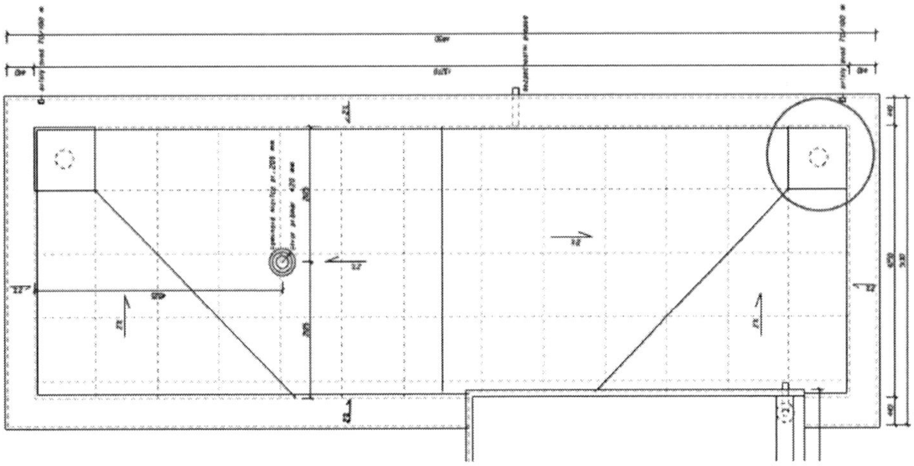

Fig. 2. Roof plan.

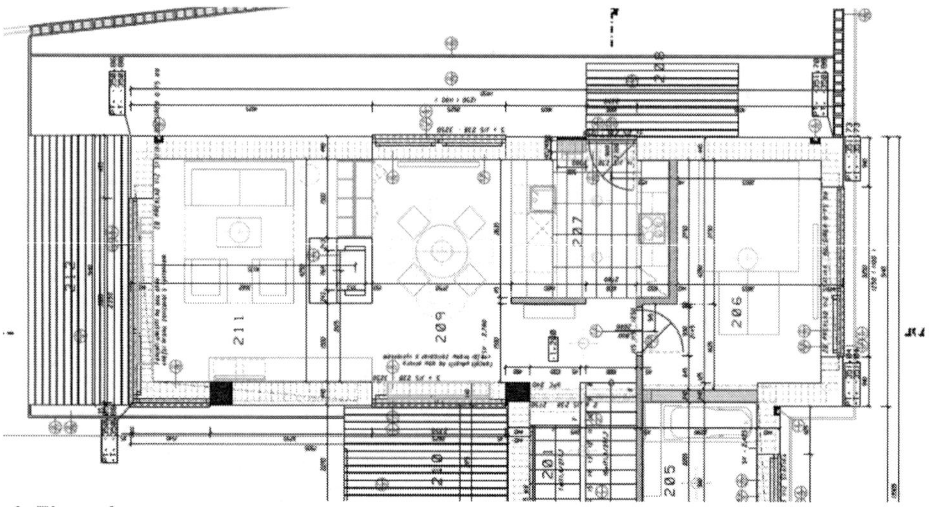

Fig. 3. Floor plan.

1.2 Construction description

Affected one-floor construction is covered by a flat roof loaded by a pea gravel drained by heated horizontal roof drains throughout the parapet wall construction into the vertical drainage situated on an outer part of a facade. Vertical construction is built from perforated blocks. Load-bearing structure of the roof consists of 160 × 220 mm beam system with OSB slab formwork where following roof composition is used:

- Loading by a pea gravel,
- mPVC hydroisolation,
- Screed to fall from heat isolation made by EPS slabs,
- Geofabric,
- Moisture barrier from light type folio,
- OSB slab framework.

2 Condensation

Water vapour condensation started to appear during season with external temperature around 0 – 4 °C with a foggy climate, or higher external humidity. Condensation was evident via humidity maps on a lower front side of OSB slab framework above the living room and bedroom. Furthermore, during the critical season humidity maps were followed by a drop leakage into the interior with capacity of 1 deciliter á 6 h.

2.1 Searching for the root cause

First the root cause might have been found in leakage of hydroisolation folio connection. During the research and inspection was possible to identify that the folio connection is in a good condition. Next a probe into the roof construction was used in order to clarify if there is any extreme water vapour and heat isolation moisture on the moisture barrier interface. The probe did not give any proof of a higher level of heat isolation moisture. On the other hand, it showed that the roof drains had not been in function and in the cabel area, which enables roof drain supply, sufficient sealing and connection to the moisture barrier was not carried out. In addition, discrepancy with design documentation was revealed when a heat isolation with noticeably higher thermal-technical properties should have been applied under the roof drain. Besides, roof draines were clogged by soil and due to insufficient screed to fall the water was kept locally in depressed area of a drainage and connection of a roof drain with horizontal internal (rainwater) drainage.

Based on these findings change of roof drains was performed as well as a sealing of cabel pass through by an aluminium tape designed for moisture barrier repair. Unfortunately, after six months condensation repeated in a similar range again.

After this condensation root cause needed to be revealed. It was necessary to consider possibility of moisture distribution in bad aired air layer going consequently above the living room, kitchen and bedroom. It was also necessary to take into account possibility of moisture appearance in cool corner due to insufficiently solved details with inadequate thermoisolation.

2.2 Measuring and diagnostics

Moisture measuring, screening by infra camera and modelling of chosen constructions in simulation software was used in order to identify the moisture source.

2.1.1 Thermografical measuring

Critical temperature is evident in the themogram. The temperature difference between an edge and corner of ceiling construction is obvious. Following temperature was recorded during border condition:
• Interior temperature: 22 °C,
• Interior relative humidity: 55%,
• Exterior temperature: +5 °C,
• Exterior relative humidity: 75%.

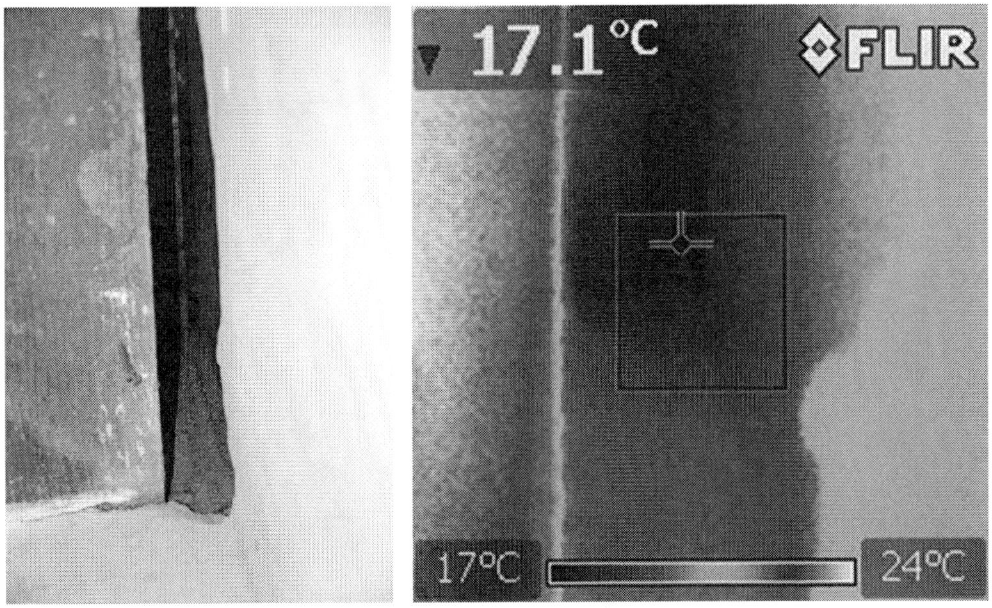

Fig. 4. Thermografical screening of the hollow on the interface of beam and circumferential construction (bedroom).

Dew point temperature is during this condition 15.992°C. The temperature was almost reached and even though the exterior border condition was really advantageous.

DOI: 10.1051/matecconf/20179301004

2.1.2 Moisture measuring

Relative humidity between the beam and view from below and temperature of few areas was measured in order to determine moisture.

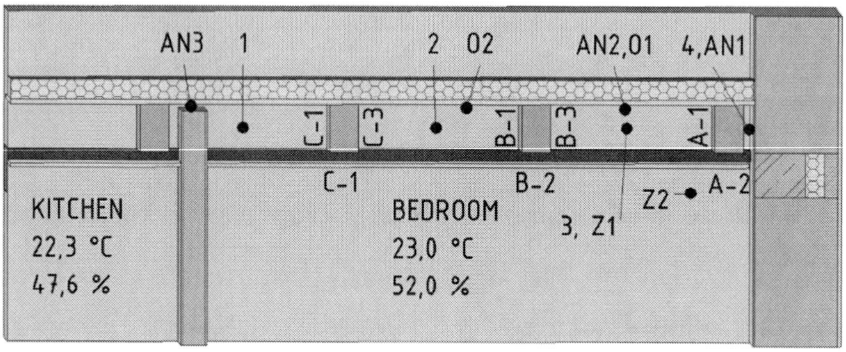

Fig. 5. Measuring points position.

Measuring was performed by MOIST device and the moisture of approximately 20–35% was shown. Based on the calibration curves, which had been specified during previous measuring, this moisture responds to the 10% spruce wood volume humidity. This can be considered as a normal moisture of build-in wood and is not a sign of significant surface condensation yet.

2.1.3 2D detail in thermal field

Actual 2D modelling of critical area during local border conditions was used. Applied isotherm represents critical temperature development. In an interruption area this requirement is underestimated which brings water condensation. Water vapour condensation is even higher in the roof area, which is weakened in the roof catch basin.

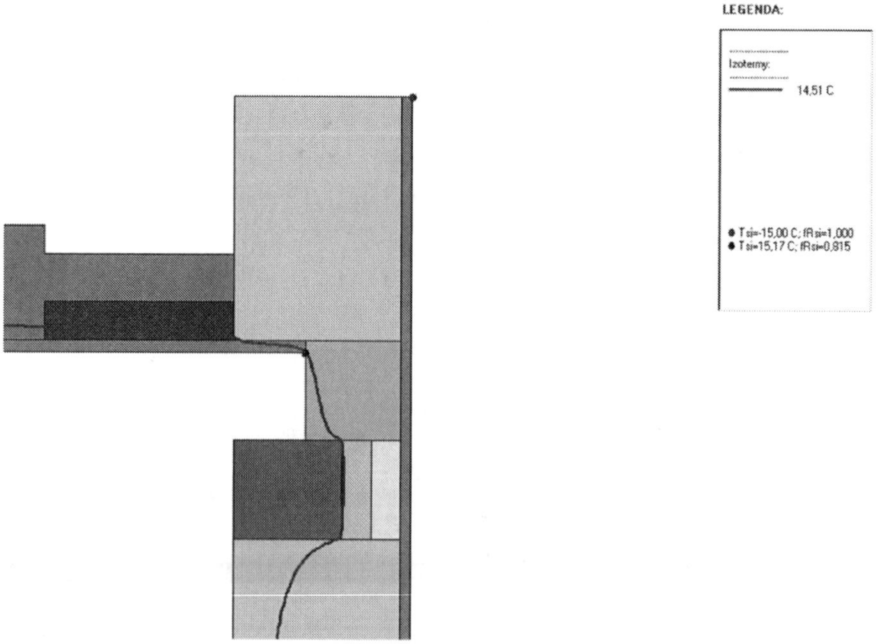

Fig. 6. Critical corner 2D modelling.

 DOI: 10.1051/matecconf/20179301004

2.1.4 Flow simulation by CFD

In order to take into consideration a moisture occurence due to a distribution in bad aired air layer a simulation of CFD (Computational Fluid Dynamics) was used to examine real flow in the building in comparison with real measuring on the site. Thanks to the simulation it was possible to exclude significant influence of redistribution of moisture from neighbouring rooms.

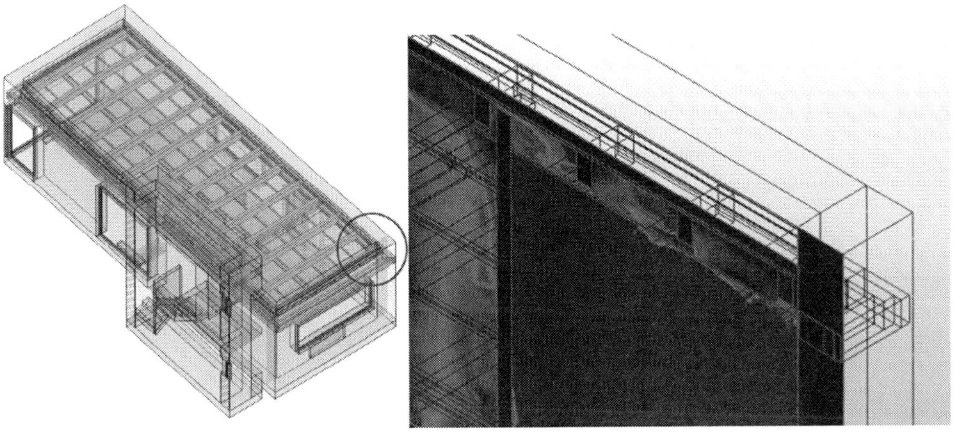

Fig. 7. Area and output example from CFD simulation.

3 Summary

Affected roof construction consists of moisture barrier designed as light type folio which is neither standard nor recommended solution as the diffusion resistance is not as high as in case of heavy asphalt membrane and it is easier to perforate this layer. Base layer of moisture barrier layer consists of OSB slab, which is not sufficiently sealed with circumferencial vertical bearing construction and thus the arisen hollow brings higher air moisture level. In this hollow high relative moisture was measured (approximately 55%) in case the view from below was opened. During its closure even higher moisture is highly to be expected.

Based on measured values, performed simulations and actual situation analysis it is obvious that the condensation and leakage problems are caused by weakening (lowering of thickness) of themoisolation building cover in roof drainage area. The main function of this isolation is to minimize thermal bridge occurance and decrease surface condensation risk. Weakening has significant influence on surface temperature as well as surface condensation occurance. This impact is multiplied by the fact that under the roof drainage a thermal isolation with low thermal conductivity coefficient was chosen.

Damage of moister barrier layer and in general insufficient detail conception of its connection may negatively contribute to higher condensation inside of the construction in drainage area and then condensate return into the interior. These high-risk areas appear in this building fully logically near drainage fastening, alternatively in connection of the moisture barrier to parapet wall.

References

1. *ČSN EN ISO 10211 Thermal bridges in building construction - Heat flows and surface temperatures - Detailed calculations* (2009)
2. *ČSN 73 0540-2 Thermal protection of buildings - Part 2: Requirements (2011)*
3. *ČSN 73 0540-3 Thermal protection of buildings - Part 3: Design value quantities (2005)*
4. K. Chaloupka, Z. Svoboda, *Ploché střechy: Praktický průvodce* (Prague, Grada Publishing, 2009)

Calculation and reliability of timber structure bolt connection subjected to double unequal shears with thick plates as outer members

Josef Musilek[1,*] and *Luboš* Podolka[1]

[1]Institute of Technology and Business in České Budějovice, Department of Civil Engineering, Okružní 517/10, 370 01 České Budějovice, Czech Republic

Abstract. The paper shows the calculation and reliability assessment of bolted connections subjected to unequal double shear with thick plates as outer members and inner timber member. This situation can occur very often during the designing of the timber structure bolt connection. Bolted connections subjected to unequal double shear are not described in present European standards where there is described only calculation and reliability of timber structure bolt connection subjected to double equal shears. So structure engineers have no rules which help them during a designing process when they are confronted with this case. The concrete example for the case of unequal shears is calculated and explained.

1 Introduction

In [1] and [2] and there were derived formulas for determination of design load-carrying capacity and for assessment of reliability in ultimate limit state of bolt connections subjected to unequal double shears with thick plates as outer members and inner timber member. In present there is no method contained in [3] which would give a procedure how to calculate design load-carrying capacity, and the reliability in ultimate limit state of such connection.

2 Example of connection

The connection arrangement according to is assumed.

* Corresponding author: musilek@mail.vstecb.cz

 DOI: 10.1051/matecconf/20179301005

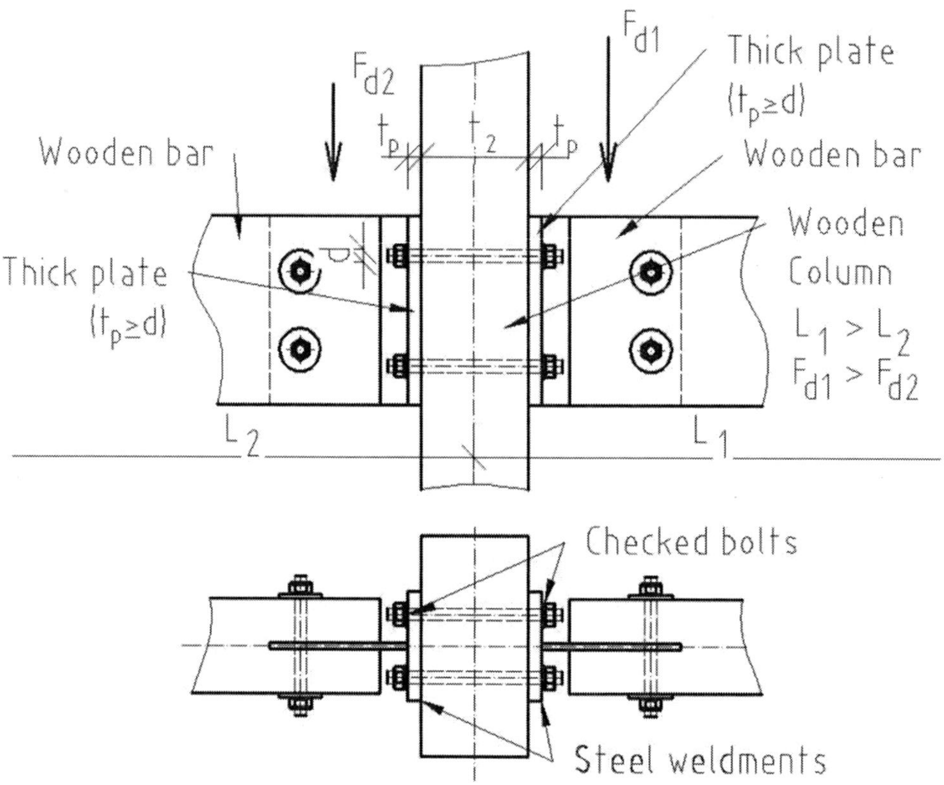

Fig. 1. Connection arrangement.

Fig. 1 presents a connection of wooden bars with a wooden column. The connection between the bars and the column is made by steel elements of T shape and by bolts. The spans of bars L_1 and L_2 are not equal, $L_1 > L_2$ so the design load F_{d1} is bigger than F_{d2}. The loads F_{d1} and F_{d2} strain the bolts in girder by shear. The plates at the girder are classified as thick plates, because $t_p \geq d$, and it is assumed that the tolerance allowance for the bolt hole is less than $0,1d$. It is also assumed, that the materials of the connection is not degraded by biological pest for example as it is shown in [4].

In the example, there is only checked the shear resistance of the bolts in the column. These bolts are marked in as "Checked bolts".

The input variables are as follows:

- design loads of the connection: $F_{d1} = 32kN$; $F_{d2} = 12kN$,
- width of the middle timber part: $t_2=60mm$,
- material of the middle timber part: C24,
- apparent density: $\rho_k=370\ kg/m^3$,
- angle between the load directions and the grain of the middle timber part: $\alpha = 0°$,
- characteristic tensile strenght of the bolt: $f_{uk} = 360MPa$,
- diameter of the bolts: $d = 16mm$,
- tensile bolt area: $A_s = 157mm^2$,
- thickness of the outer steel plates: $t_{plate} = 16mm$,
- modification factor: $k_{mod} = 0,8$,
- partial factor for timber: $\gamma_M = 1,3$.

The design load-carrying capacity for different failure modes were derived in [1] and [2]. When we increase the loads in the way that the ratio $n=F_{d2}/F_{d1}$ stays constant, the following formulas for the design load-carrying capacity are as follows:

Failure mode a) (the member for characteristic withdrawal capacity of the bolt according to [3] is omitted)

$$R_{d1} = \frac{k_{mod}}{\gamma_M}(2{,}3 \cdot \sqrt{M_{y,Rk} \cdot f_{h,2,k} \cdot d}) \tag{1}$$

Failure mode b)

$$R_{d1} = \frac{k_{mod}}{\gamma_M} \cdot \frac{t_2 \cdot f_{h,2,k} \cdot d}{1+n} \tag{2}$$

Failure mode c)

$$R_{d1} = \frac{k_{mod}}{\gamma_M} \cdot f_{h,2,k} \cdot d \cdot \frac{\sqrt{{t_2}^2 \cdot 2 \cdot (1+n^2) + (1+n)^2 \cdot \frac{8 \cdot M_{yk}}{f_{h,2,k} \cdot d}} + t_2 \cdot (n-1)}{(1+n)^2} \tag{3}$$

At first we calculate some values needed for further calculation:

Characteristic embedment strength in the hole for the bolt in the girder $f_{h,2,k}$:

$$f_{h,0,k} = 0{,}082 \cdot (1 - 0{,}01 \cdot d) \cdot \rho_k = 0{,}082 \cdot (1 - 0{,}01 \cdot 16) \cdot 370 = 25{,}49MPa \tag{4}$$

$$k_{90} = 1{,}35 + 0{,}015 \cdot d = 1{,}35 + 0{,}015 \cdot 16 = 1{,}59 \tag{5}$$

$$f_{h,2,k} = \frac{f_{h,0,k}}{k_{90} \cdot (\sin\alpha)^2 + (\cos\alpha)^2} = \frac{25{,}49}{1{,}59 \cdot (\sin 0)^2 + (\cos 0)^2} = 25{,}49MPa \tag{6}$$

Characteristic bolt yield moment

$$M_{y,Rk} = 0{,}3 \cdot f_{u,k} \cdot d^{2,6} = 0{,}3 \cdot 360 \cdot 16^{2,6} = 1{,}459 \cdot 10^5 Nmm \tag{7}$$

Ratio n:

$$n = \frac{\mathrm{F_{d2}}}{\mathrm{F_{d1}}} = \frac{12}{32} = 0{,}375 \tag{8}$$

Design load-carrying capacity for the failure mode a)

$$R_{d1} = \frac{k_{mod}}{\gamma_M}\left(2{,}3 \cdot \sqrt{M_{y,Rk} \cdot f_{h,2,k} \cdot d}\right) = \frac{0{,}8}{1{,}3} \cdot \left(2{,}3 \cdot \sqrt{1{,}459 \cdot 10^5 \cdot 25{,}49 \cdot 16}\right) = \\ = 10{,}92kN \tag{9}$$

Design load-carrying capacity for the failure mode b)

$$R_{d1} = \frac{k_{mod}}{\gamma_M} \cdot \frac{t_2 \cdot f_{h,2,k} \cdot d}{1+n} = \frac{0{,}8}{1{,}3} \cdot \frac{60 \cdot 25{,}49 \cdot 16}{1+0{,}375} = 10{,}95 kN \tag{10}$$

Design load-carrying capacity for the failure mode c)

$$R_{d1} = \frac{k_{mod}}{\gamma_M} \cdot f_{h,2,k} \cdot d \cdot \frac{\sqrt{t_2{}^2 \cdot 2 \cdot (1+n^2) + (1+n)^2 \cdot \frac{8 \cdot M_{yk}}{f_{h,2,k} \cdot d}} + t_2 \cdot (n-1)}{(1+n)^2}$$
$$= \frac{0{,}8}{1{,}3} \cdot 25{,}49 \cdot 16 \cdot \frac{\sqrt{60^2 \cdot 2 \cdot (1+0{,}375^2) + (1+0{,}375\ )^2 \cdot \frac{8 \cdot 1{,}459 \cdot 10^5}{25{,}49 \cdot 16}} + 60 \cdot (0{,}375 - 1)}{(1+0{,}375)^2}$$
$$= 10{,}52\ kN \tag{11}$$

Resultant design load-carrying capacity:

$$R_d = \min(10{,}92kN, 10{,}95kN, 10{,}52\ kN) = 10{,}52\text{kN} \tag{12}$$

The reliability:

$$n = \frac{\mathrm{F_{d1}}}{4 \cdot \mathrm{R_d}} = \frac{32}{4 \cdot 10{,}52} = 0{,}760 \tag{13}$$

There is also possibility that the load F_{d2} remains constant and the load F_{d1} is increasing only. So we can write $R_{d2} = F_{d2} / 4$. In [1] and [2] formulas were derived for this case. If we use them for our example, we obtain the following results:

Failure mode a) (the member for characteristic withdrawal capacity of the bolt according to [3] is omitted)

$$R_{d1} = \frac{k_{mod}}{\gamma_M}\left(2{,}3 \cdot \sqrt{M_{y,Rk} \cdot f_{h,2,k} \cdot d}\right) = \frac{0{,}8}{1{,}3} \cdot \left(2{,}3 \cdot \sqrt{1{,}459 \cdot 10^5 \cdot 25{,}49 \cdot 16}\right) =$$
$$= 10{,}92kN \tag{14}$$

Failure mode b)

$$R_{d1} = \frac{k_{mod}}{\gamma_M} \cdot f_{h,2,k} \cdot d \cdot t_2 - R_{d2} = \frac{0{,}8}{1{,}3} \cdot 25{,}49 \cdot 16 \cdot 60 - \frac{12000}{4} = 11{,}47\ kN \tag{15}$$

Failure mode c)

$$R_{d1}=\frac{k_{mod}}{\gamma_M}\cdot f_{h,2,k}\cdot d\cdot\left(\sqrt{2\cdot\left(t_2{}^2+\frac{4\cdot M_{yk}}{f_{h,2,k}\cdot d}+\frac{2\cdot\gamma_M\cdot R_{d2}\cdot t_2}{k_{mod}\cdot f_{h,2,k}\cdot d}\right)}-t_2\right)-R_{d2}$$

$$=\frac{0{,}8}{1{,}3}\cdot 25{,}49\cdot 16\cdot\left(\sqrt{2\cdot\left(60^2+\frac{4\cdot 1{,}459\cdot 10^5}{25{,}49\cdot 16}+\frac{2\cdot 1{,}3\cdot\frac{12000}{4}\cdot 60}{0{,}8\cdot 25{,}49\cdot 16}\right)}-60\right)-\frac{12000}{4}$$

$$=10{,}48kN \tag{16}$$

Resultant design load-carrying capacity:

$$R_d=\min(10{,}92kN, 11{,}47kN, 10{,}48\,kN)=10{,}48\mathrm{kN} \tag{17}$$

The reliability:

$$n=\frac{\mathrm{F_{d1}}}{4\cdot\mathrm{R_d}}=\frac{32}{4\cdot 10{,}48}=0{,}7 \tag{18}$$

3 Summary

The example for reliability assessment of bolt connection of bolted connections subjected to unequal double shear with thick plates as outer members and inner timber member was presented. The two different ways of loading were considered. The differences between these two ways of loading were not significant for the reliability assessment.

References

1. K. Kubečka, J. Musílek, *Contemporary Problems of Architecture and Construction* (Univerzita Degli Study Firenze, 7, 59-62, 2015)
2. J. Musílek, J. Žák, *8th International Conference Building Defects* (The Institute of Technology and Business in České Budějovice, , České Budějovice, to be published)
3. *EN 1995-1-1, Design of timber structures. Part 1-1: General-Common rules and rules for buildings (1995)*
4. P. Bednářová, A. Hynková, Littera Scripta (The Institute of Technology and Business in České Budějovice, České Budějovice, 2009)

Redevelopment of the roof structure in the production plant of Skoda Auto a.s. in Kvasiny

Luboš Podolka[1,*]

[1]University of technology and economics in Czech Budejovice, Department of civil engineering, 37001 Czech Budejovice, Czech Republic

Abstract. The paper describes a proposal for the redevelopment of the roof structure using FRP materials. During the design of the redevelopment, the experience of experiments with FRP materials that took place in the PAVUS a.s. laboratory in Veselí nad Lužnicí and laboratory of Faculty of Civil Engineering of the Czech Technical University in Prague was used. Initial experiments used for the later proposal of redeveloping supporting structures are also described and some are compared with the mathematical model of the ATENA software. The experiments in the PAVUS laboratory resulting in the adjustment of the anchoring length correspond to the required fire resistance. The result of the Faculty of Civil Engineering of the Czech Technical University in Prague laboratory experiment is the design of the optimal shape of the FRP shear reinforcements. In the article there are also photos from the implementation of the roof redevelopment.

1 Introduction – Description of reinforces structure

Part of the assembly hall was built in the second half of 2005 as an annex for the Roomster model production line.

The existing load bearing structure of the assembly hall annex consists of a prefabricated reinforced concrete skeleton hall with a width of naves 2 x 18 m and a zone located by the northern facade with a built-in other floor with a span of supporting structures measuring 6.0 m. In the longitudinal direction, the hall consists of three separate dilatation parts with a column module measuring 12.0 m. The headroom under the trusses of the hall is about 9.5 m.

A part of the building has two floors. The vertical structure of the assembled skeleton of the hall consists of columns with a square (or rectangular) cross-section with a side length of 600 mm.

The roof structure has the shape of a gabled roof with a gradient of 2% and inner valleys. On a major portion of the roof there are transverse separated arc-shaped skylight strips made of Makrolon – the purlins of the gable shape support the transverse (in the direction of the skylights) beams along skylights, which then support a pair of prefabricated concrete skylight rims. The roof cladding is lightweight and insulated with a supporting

* Corresponding author: podolka@mail.vstecb.cz

DOI: 10.1051/matecconf/20179301006

structure made of a trapezoidal sheet, on which is a roof waterproofing membrane, stressed due to wind suction load, anchored by self-tapping screws (through a soft thermal insulation layer).

Trusses (beams) have a longitudinal direction span of 12.0 m. In terms of load transfer, they function as continuous beams of six, or four fields with embedded joints in approximately a quarter of the range field. Statically it is a continuous beam with internal joints (i.e. Gerber girder). Trusses form the gabled roof. Their span is 18.0 meters and the axial distance between the trusses is 6.0 meters. In the trusses there are circular holes in the web for the passage of wiring.

Trusses (beams) have a reinforced concrete T-shaped cross section with the web measuring a height of 1.05 m and a width of 220 mm. The upper flange has a width of 400 mm and height of 200 mm - the haunch between the web and the flange is made with a height of 50 mm. Purlins with a height of 1.45 m are designed as garble concrete elements with a T-shaped cross section with a slope of 2%. In their web with a width of 160 mm there are circular holes for the passage of wiring. To the top flange with a width of 360 mm (in the field with a height of 150 mm and placement at a height of 300 mm) there is anchored a trapezoidal sheet with a height of 150 mm, which forms a supporting structure of the roofing.

2 Condition of supporting roofing structure

The existing concrete structure (especially prefabricated trusses in modular axis "F") is damaged mainly by shear cracking caused by static load effects (width of cracks caused by bending stress is significantly smaller). According to a more accurate measurement by optical micrometer and inaccurate measurement by triangular scale the width of cracks varies from 0.4 to 2.5 mm (in one case even 3.5 to 4.0 mm).

2.1 Based on the findings of the structural and technical survey of selected elements of the prefabricated roof parts it can be stated as follows

- The profile and location of the built concrete reinforcement practically corresponds with the production documentation,
- The strength grade of the concrete in trusses (beams) was determined as class C 20/25 (B 25) and does not conform to the required 500 mark of concrete recorded in the drawings (mark of the required concrete strength corresponds with class C 35/45),
- Modulus of concrete elasticity in the compression of collected samples was determined as 23.8 GPa.

2.2 Based on the results of comparative calculations for the selected elements of the prefabricated roof parts it can be stated as follows

- The shear resistance of the reinforced concrete sections for trusses (beams) is due to inappropriately spaced stirrups, insufficient for practically their entire length,
- Areas of discontinuities of reinforced concrete cross-sections (webs weakened by circular holes and changes in the cross-sectional height of placement on lock joints) are not sufficient due to the lack of embedded tensile reinforcement.

Fig. 1. Reinforced roof structure.

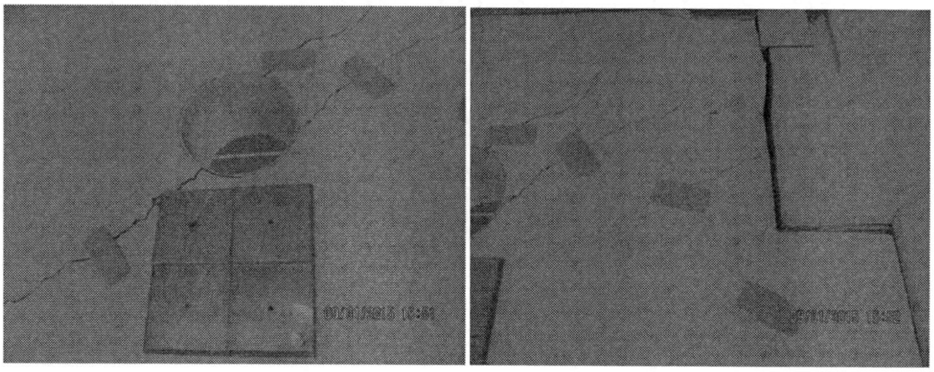

Fig. 2. Details of shear cracks in the beam.

3 Draft remediation design

The redevelopment of the structure is designed using FRP materials in the form of shear and bending reinforcement, FRP fabrics. The assessment of cross-sections and design of reinforcement is based on the principle of truss analogy that describes the behavior of the structure girder in the place of the mutual placing of elements on each other, where they form short bracket joints of a pair of beams [1-2].

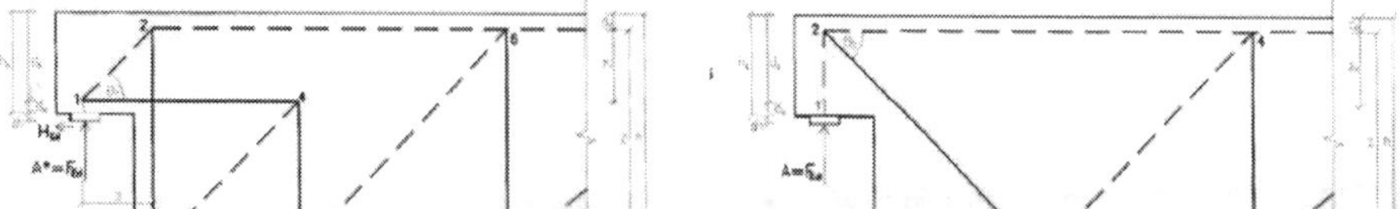

Fig. 3. Models of placing truss based on truss analogy.

FRP material contribution to shear resistance, V_{fd}, "reinforcement by vertical fabric"

For example: Sheet 240, E_{fu} = 240 GPa, thickness t_f = 1,176.10-4 m, $\varepsilon_{fd,e}$ = 0.002 up to 0.004, at angle $\alpha = 90°$, i.e. vertically.

$$V_{Rd} = \min\left(V_{cd} + V_{wd} + V_{fd}, V_{Rd2}\right) \tag{1}$$

where the bearing capacity of the FRP reinforcement is calculated using following formula:

$$V_{fd} = 0.9.\varepsilon_{fd,e}.E_{fu}.\rho_f.b_w.d.(\cot\theta + \cot\alpha).\sin\alpha \tag{2}$$

$\varepsilon_{fd,e}$ ... a design parameter of effective FRP deformation
b_w ... minimal cross-sectional width above the effective depth
d ...effective cross-sectional height
ρ_f ... *FRP,* reinforced, straight $2t_f.\sin\alpha/b_w$ for the bevel cut, shear reinforcement thickness t_f
(b_w = the minimum width of the concrete cross section above effective height),
or $(2t_f/b_w).(b_f/s_f)$ for FRP reinforcement in the form of strips of fabric with width b_f at the span s_f
E_{fu} ... elasticity modulus of FRP in the direction of fiber orientation
θ ... angle of diagonal shear cracks from the plane, mostly 45°
α ... angle between the fibers and the plane

4 Experimental verification of anchoring length slats

To determine the anchor length of FRP materials there were created divided bars with the dimensions 150 x 150 x 600 mm with an embedded reinforcement Ø20mm provided with an anchor plate sheet thickness = 8 mm, dimensions 150x150 mm. The reinforcement was butt-welded to the plate. On the outer face of the bars there were glued slats with different anchor length; 50 mm, 100 mm and 150 mm from various concretes: C 16/20, C 30/37 and C 50/60. To verify the effect of the temperature on the anchorage length, tests were conducted in the PAVUS laboratory on bars measuring 150 x 150 x 800 mm, with 500 mm slat, which was unprotected or protected against higher temperatures using cladding from the Grena a.s. company, boards made from Vermiculite.

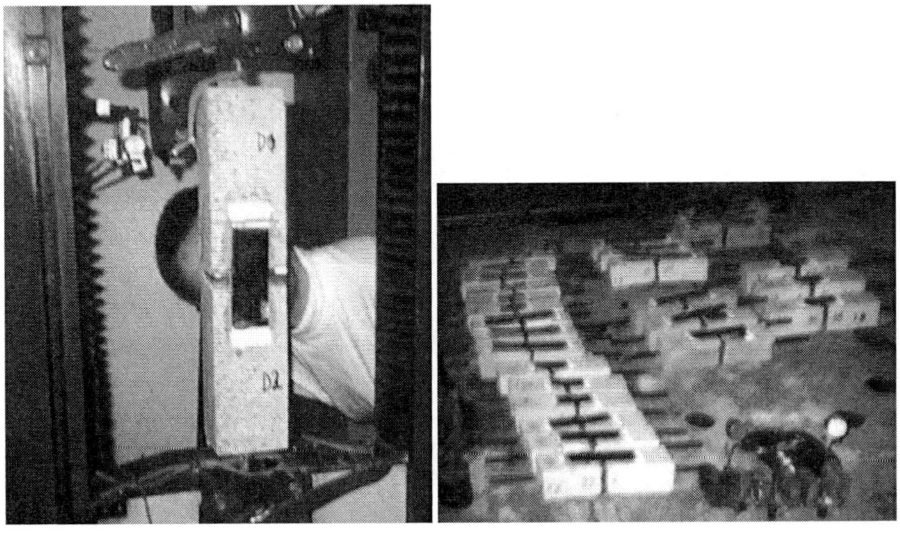

Fig. 4. Experimental tests to verify the length of anchor slats

DOI: 10.1051/matecconf/20179301006

Fig. 5. Preparation of an experimental test in a fire test furnace. Left fixing cladding on glued slats, on the right finished cladding with two slats left without protection.

Fig. 6. Slats after fireproof cladding was removed (left), on the right a detail of sawn slats and exposed concrete.

The experiment determined the requirement of extending the length of the anchoring slats by 50 mm in order to ensure sufficient fire resistance in comparison with an anchor length without requirement for fire resistance.

The alternative to facing boards of Vermiculite for FRP plates for FRP lining fabric using a fiberglass covered with cement plaster.

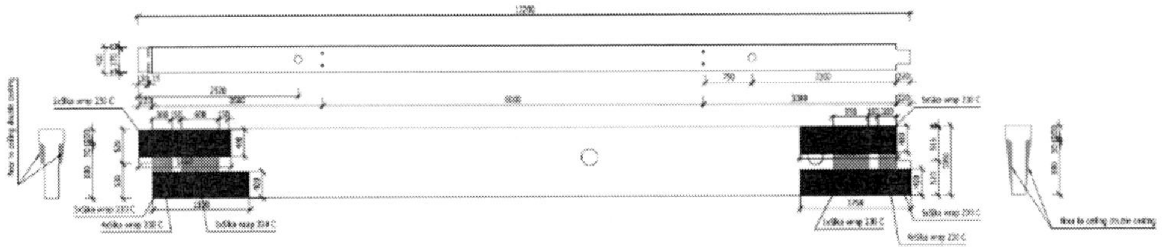

Fig. 7. Shape of the truss, its reinforcement and the proposed reinforcement of the truss using FRP fabrics.

Fig. 8. Reinforcement using FRP fabric.

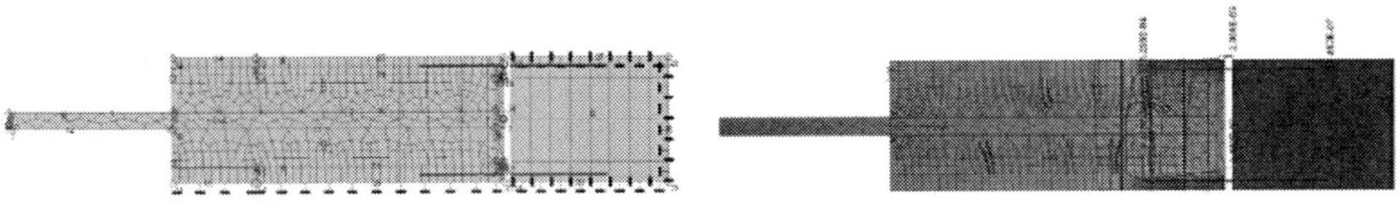

Fig. 9. The mathematical model to verify the anchor length created in the Atena software and The stress in the cross section shown in the mathematical model.

The above described experimental element was modeled in program ATENA, so that it can monitor the behavior of the element in place both active and passive anchors, i.e. power transfer from reinforcing bars through mechanical anchoring to the concrete and spread stresses in the concrete for anchors.

5 Conclusion

The commenced assembly of a new car model at the Skoda Auto a.s. factory in Kvasiny proved the functionality of the proposed solution. Since the application of the reinforcement there has not been any damage to the roof structure caused by the applied load.

References

1. *EN 1992-1-1 Eurocode 2: Design of concrete structures. Part 1-1: General rules and rules for buildings* (1992)
2. *Externally bonded FRP reinforcement for RC structures Technical report on the "Design and use of externally bonded fiber reinforced polymer reinforcement (FRP EBR) for reinforced concrete structures" by 'EBR' working party of fib TG 9.3* (2001)

MATEC Web of Conferences **93**, 01007 (2017) DOI: 10.1051/matecconf/20179301007
Building Defects 2016

Analysis of the causes of defects due to quality roofing membranes mechanical anchoring

Jindřich Stodůlka[1,*], *Markéta* Bogárová[1], and *Libor* Matějka[1]

[1]VUT FAST, Ústav pozemního stavitelství, Veveří 331/95, 602 00 Brno, Czech Republic

Abstract. Anchoring of the layers of flat warm roofs using anchoring elements is nowadays one of the most used technology. This is a frequent technology to repair damaged roofs (total reconstruction, increasing the load on the wind loads, partial repairs etc.). In the European building market there are several suppliers of the enchoring equipment who declare their product's properties in the european technical approvements (ETA). The goal of the article is to demonstrate results of the carried out measurements and their possible impact into practical application of the mechanically anchored systems. This allows reduce the amount of defects to improving flexible watertight membranes. The topic can be considered actual and due to the large number of defects precisely for these types of roofs.

1 Introduction

Based on the issue of the Directive ETAG 006 [1] are all anchoring elements tested in accordance with unified methodology. The directive ETAG 006 [1] is a document the serves as a basis to test and design the mechanically anchored roofs. Development of anchoring elements with telescopes is combined with growing demands to reduce the impact of spot thermal bridges that arise due to anchoring. [2]

Quality, quality and design of the telescope has a significant influence on the resistance and therefore the durability and longevity of flat roofs. Poorly made plastic telescopes are often the cause of the crash roofs. [3]

ETAG 006 [1] defines the roof composition as a system formed with cover waterproofing, thermal insulation, anchoring element and load-bearing layer. Many times is the system defined with the MEFAWAME abbreviation from English "Mechanically fastened waterproofing membranes".

* Corresponding author: stodulkaj@study.vutbr.cz

2 Types of tests and application of their results in the mechanic analysis and review

The directive recognizes 3 basic types of test, delivering data for the following mechanic analysis of the anchoring (anchoring plan). The basic tests are: large scale (real) test, small scale tests and extracting tests in situ.

The principal of all of the above stated test is to gain information about given system or its component, which are needed for definition of the maximum allowed design loading on one anchoring element – Wadm [N] value.

2.1 The real scale test

The real scale test are based on testing of the all roof system – the cover waterproofing, anchoring elements, thermal insulation and load-bearing layer. Full segment of a roof composition is tested in so called pressure chamber that simulates the increasing loading with wind suction. The intensity and increasing of the wind suction are thoroughly described in the directive EAG 006. The result is to define design resistance of one anchoring element Wadm [N] for the concrete system with concrete components from given producers or suppliers of the waterproofing covers and anchoring material.

If there are used system that are tested with the real scale test in the roof, this design of the system is considered the most reliable. According to ČSN EN 1990 [4] the results of the real scale test are the only valid indicators of the suitability of usage of MEFAWAME for its design. However currently the results of the large scale test themselves are possessed only by few of the waterproofing covers. The main portion of the suppliers of the waterproofing covers still declares usual value 400N per one anchoring element.

2.2 Extraction tests in situ

The extraction test are carried out in cases when the loading capacity of the real load-bearing layer is not sure. The aim of the extraction test is to verify behavior of the concrete anchoring element with the concrete load-bearing layer present in the structure. Output of the tests is to define design loading per one anchoring element but considering the real scale test results.

It is necessary to obtain results from minimum of six test per 5000 m^2 of the roof area evenly throughout the roof plane. Minimally 50% from the total number of the test must be carried out in the area of corners and roof edges.

2.3 Small scale tests

The methodology of the MEFAWAME in accordance with directive ETAG 006 allows interchange of some parts or execution of the tested MEFAWAME compared to the real scale test. For such cases is used result of the small scale test which is carried out on each of the roof system component separately. It is possible to swap p.e. anchoring element, cover waterproofing, load-bearing layer etc. The advantage of this types of tests compared to the real scale tests is significantly lower cost.

To allow such interchange in previously tested system (by the real scale test) it is necessary to know the properties of the elements which are interchanged within MEFAWAME. Even here the directive specifies testing of individuals components of MEFAWAME – these tests are considered small scale tests. The output of the small scale test is a correlation coefficient k [-], which enters the calculation of the Wadm value according to relation 1 [1].

$$W\_(adm,nc)=k\cdot W\_(adm,oc) \text{ [N]} \quad (1)$$

The Wadm,oc [N] is a design resistance per one anchoring elements of the non-swapped original (from English "Old combination) system MEFAWAME, which is after multiplication with the correlation coefficient k[-] in accordance with the relation 1 leads to the definition of the design resistance of one anchoring element new (from English. „New Combination") system MEFAWAME.

In anchoring elements the correlation coefficient is defined as a division of loading capacity of the new and original anchoring element – formula 2 [1].

$$k(-)=R_{nc}(N)/R_{oc}(N) \quad (2)$$

The values R_{nc} [N] and R_{oc}[N] are the results of the tension loading capacity of the anchoring elements combined with an exact load-bearing layer. These values are obtained from the results of the small scale tests and the anchoring material suppliers state those in their ETA's. In correlation coefficient which enters the formula for the Wadm,nc calculation there is a restriction for it to be within the interval 0,5 – 1,0.

3 Experimental measurement of the axial tension loading capacity of the anchoring elements.

Experimental measurement of the axial tension loading capacity was assembled using the „Axial loading test" criteria which is precisely defined in ETAG 006 annex D [1]. From the EN 1990 [4] standard point of view the carried out experimental measurement can be considered a "inspection test to verify identity or quality of given elements" (ČSN EN 1990 - annex D.3 letter. e). As a test samples were chosen anchoring elements with telescopes of length 85 – 105 mm, that were added with screws 44,8 x 60 into trapezoidal metal sheeting.

There were chosen samples from 9 suppliers of the anchoring material from the European building market while every test sample was represented with 10 test bodies. The test bodies were tested in combination with load-bearing layer formed with trapezoidal metal sheeting S280GD of the thickness 0,75 mm. The sheeting S280GD was chosen on purpose because it is mostly used load-bearing layer especially in industrial buildings and shopping center buildings.

3.1 Testing device and test bodies

The tearing machine Heckert FT10-1, which was wired to a desktop computer equipped with the CatmanEasy software tool was used for the experimental measurement. During the experiment there were recorded actual forces, displacement and time. The loading speed was set to 10 mm per minute.

The test bodies were formed with anchoring elements fixed into a metal sheet of 100 x 100 mm. The anchoring was carried out using accumulator screwdriver without pre-drilling. In the next step the elements were placed into welded elements securing axial transmission of the force into the anchoring element pre-stressed with 5 N and tested. The graphical scheme of the test is shown in the Fig. 1 and. Fig 2.

 DOI: 10.1051/matecconf/20179301007

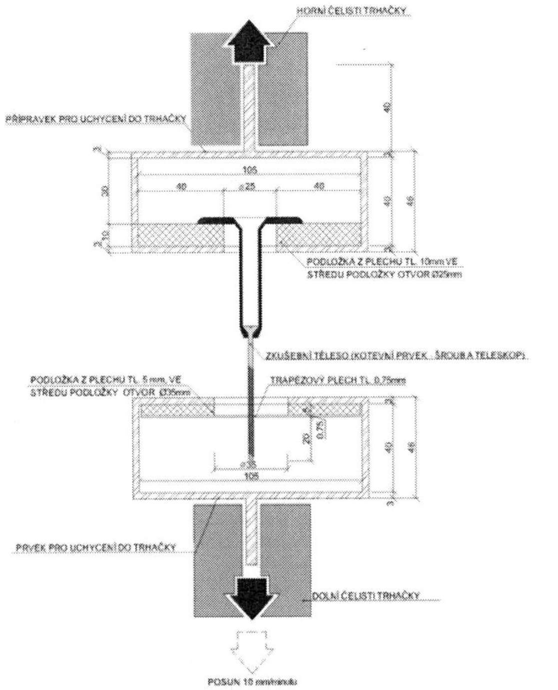

Fig. 1. The scheme of the axial tension loading capacity of anchoring elements test.

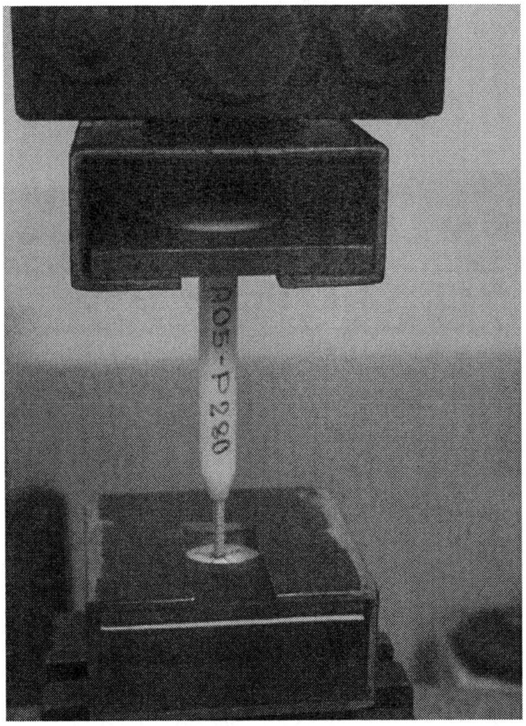

Fig. 2. The test body attached to the testing device.

Pre-stressing to the value of 5 KN was carried out to secure identical initial position in all tested bodies.

3.2 Results of experiments, relative and absolute comparion of the middle and characteristic values of loading capacity of anchoring elements

The input data for the mechanical analysis was the values of maximal forces F [N], under which the failure in loading capacity of the test bodies occurred. The characteristic values were defined in accordance with the procedure prescribed in ETAG 006 [1] and ČSN EN 1990 – annex D [4]. The aim was to compare individual test samples with each other and to gain data on the quality of the anchoring elements of the individual suppliers of the anchoring material. Next it was aimed to compare measured values with the values stated in the supplier's ETA's The results of the individual test samples are grouped in the Table 1.

Table 1. Relative comparison of loading capacity of anchoring elements with telescopes.

No.	$F_{average}$ [N]	$F_{k,n}$ [N]	$F_{k,l}$ [N]	$F_{K,ETA}$ [N]	No.	$F_{average}$ [N]	$F_{k,n}$ [N]	$F_{k,l}$ [N]	$F_{K,ETA}$ [N]
A	1458	1396	1398	1060	**G**	1361	1309	1309	525
B	1379	1312	1313	997	**H**	1442	1403	1403	1060
C	1422	1348	1349	1680	**I**	1414	1348	1349	1430
D	1329	1273	1274	890	**J**	1417	1324	1326	900
E	1396	1356	1356	1150	**K**	1454	1376	1378	1430
F	1380	1295	1297	970					

$F_{average}$ average force during loading capacity loss in anchoring element
$F_{k,n}$ characteristic loading capacity of anchoring element in normal scatter,
$F_{k,l}$ characteristic loading capacity of anchoring element in log-normal scatter,
$F_{k,ETA}$ characteristic loading capacity declared by the manufacturer in ETA.

From the absolute comparison of the characteristic values is visible that in some test samples the measured values does not reach the values stated by the manufacturers in their ETA's. By comparing of the characteristic values in the test sample "I" and "K" it is shown the there is an obvious similarity of the measured middle values of loading capacity $F_{average}$ and values $F_{k,ETA}$, that are declared by the manufacturer as the characteristic ones. In the samle "G" the declared values is on contrary significantly lower than the results of the measured values. After precise evaluation of the declared value $F_{k,ETA}$ there was observed similarity with the value $F_{O,Rk}[N]$, which can be calculated using theoretical procedure acc to [5].

4 Conclusion

The experimental measurement confirmed the doubts of a correctness of a certain data declared in the ETA products. The results show that from the overall number of 11 experimental samples of products the minimum of 3 did not reach the prescribed values.

The misleadingly declared values $F_{k,ETA}$ lead to errors in the correlation coefficients k (relation 2) and therefore to the wrong definition of the Wadm,nc value. The rest of the sample products showed values in accordance with the stated ones.

The article was developed within the project no. FAST-S-16-3345 "Verifikace chování a rozvoj konstrukcí a konstrukčních prvků na bázi dřeva a kombinovaných konstrukcí ze dřeva a železobetonu" at the Faculty of Civil Engineering VUT in Brno.

References

1. *ETAG 006. Directive for European Technical Approvals – guideline for European technical approval of systems of mechanically fastened flexible roof waterproofing membranes* (European Organisation for Technical Approvals – EOTA, Brussels, 2012)
2. V. Petránek, R. Šubrt, J. Plachý, L. Nevrivová, T. Petříček, L. Kalousek, Z. Caha, AMR **732-733**, 182-185 (2013)
3. J. Plachý, J. Beránek, Střechy-fasády-izolace, 28-29 (Nakladatelství Mise, Ostrava – Vítkovice, 2013)
4. *ČSN EN 1990. Eurocode: Eurocode: Basis of structural design* (ÚNMZ, Prague, 2011)
5. *ČSN EN 1993-1-3. Eurocode 3: Design of steel structures - Part 1-3: General rules - Supplementary rules for cold-formed members and sheeting* (ÚNMZ, Prague, 2008)

Use of EMW radiation in the building industry at defects in buildings

Jindřich Sobotka [1,*] and *Radim Smolka* [1]

[1] Brno University of Technology, Faculty of Civil Engineering, Institute of Building Structures, Veveří 95, 602 00 Brno, Czech Republic

Abstract. This paper discusses theory and application of microwave radiation and experimental optimization of microwave radiation to eliminate moisture content in wood elements. It will be appreciated that the rising moisture leaking into the structure, resulting in defects and structures of the buildings themselves. Owing to its properties, microwave radiation has been used in the construction industry in modern times, in particular to dry wet masonry of buildings. Effects of electromagnetic radiation on building structures lead to relatively sharp decreases in moisture content from damp building structures or elements. The influence of electromagnetic radiation on building structures lead to oscillation of water molecules contained in the material, which cause a phase transformation of water into vapour. Consequently, the vapour evaporates from the moist material, thereby drying the element exposed to radiation. The article describes experiments carried out at the Faculty of Civil Engineering of the Faculty of Technology in Brno that demonstrate successful decrease of water content in building materials using microwave radiation. First, the understanding of microwave radiation will be discussed. Following an analysis of research results an optimum intensity of microwave radiation sources as well as the necessary length of the irradiation of microwave radiation have been determined with respect to the particular type of building material and the success rate of elimination of moisture.

1 Introduction

Microwaves are a term given to a part of electromagnetic radiation characterized by a wavelength ranging from 1 cm to 1 metre. In the form of waves the microwaves spread into space from the source. Worldwide, a frequency of 2.45 GHz with a corresponding wavelength of 12.2 cm has been designated for industrial purposes [1].

* Corresponding author: sobotka.j@fce.vutbr.cz

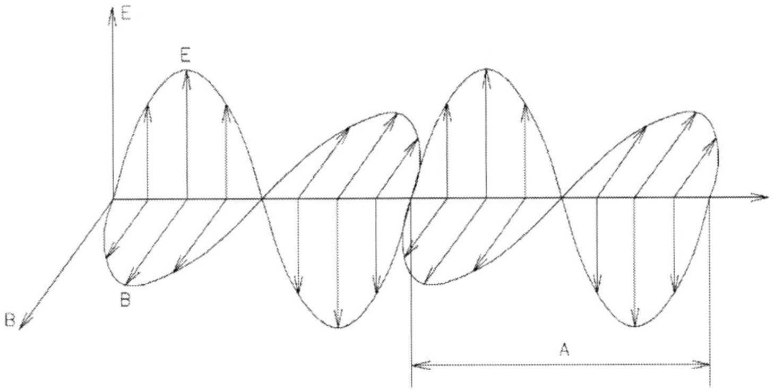

Fig. 1. Electromagnetic wave with the length λ – electric (E) and magnetic (B) components of the wave [2].

2 Drying by EMW Radiation

The water molecules although having a bi-polar character are electrically neutral Therefore, in an electric field each and every water molecule is oriented by the polarity (positive to negative and negative to positive), especially if the polarity alternates. This is precisely the case of microwave radiation. The polarity of the electromagnetic field varies depending on the frequency F, where the frequency can be higher than 10^9 times per second [1].

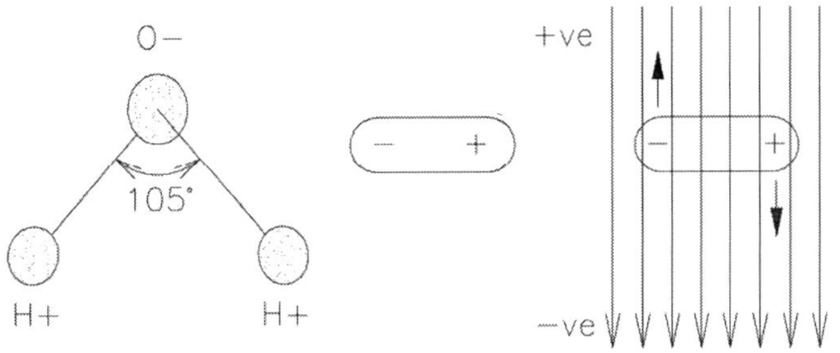

Fig. 2. Water molecule and its orientation in an electric field [2].

At high frequencies of radiation (i.e. oscillating vibration) and the subsequent change in orientation of molecules of loosely bound water produces heat. The thermal energy is produced by friction of molecules. This phenomenon is called as polar rotation or polar friction [3]. In a simple manner it can be described as follows, the entry of microwaves into the structures causes a rapid change in the polarity of molecules of loosely bound water, causing their rapid movement in the material. Because of this movement the particles themselves collide into each other, hence produce thermal energy. This process results in a conversion of water from liquid state into gaseous form, and its subsequent evaporation from the structure [4].

2.1 Advantages and disadvantages of EMW heating and radiation

The advantages of microwave radiation are:

- the so-called "Selective heating", i.e. by the radiation of multi-component structures only the materials absorbing microwave radiation are heated (i.e. materials containing loosely bound water),
- velocity of drying,

- lower energy and financial requirements in comparison with hot-air drying.

The only known disadvantage of microwave radiation application is:

- the possibility of local overheating due to inhomogeneity of the microwave field and the material itself.

In terms of interaction of building material EMW radiation there are three possible outcomes. The material is either:

- transparent – radiation passes through the material without any reaction,
- absorbing – radiation is converted to thermal energy,
- reflecting – radiation is reflected back into the environment.

For building practice the interaction of microwaves with water is the most important factor. The conversion of microwave energy into thermal energy, heat can be determined using the following relationship [1], [5].:

$$P=2\cdot\pi f\cdot\varepsilon'\cdot\varepsilon''\cdot E^2 \quad (1)$$

where: P energy absorbed per unit volume ($W\cdot m^{-1}$),
f frequency microwave field (2450 MHz),
ε′ permittivity ($F\cdot m^{-1}$),
ε′′ dielectric loss factor of the material,
E electric field intensity inside the material ($V\cdot m^{-1}$). [1]

3 Material and methods

One of a number of experiments carried out at the Faculty of Civil Engineering, Institute of Building Structures will be described in more detail. The aim was to determine the length and cyclic measurement of microwave radiation for a successful removal of moisture absorbed by wood elements [4].

3.1 Description of the experiment

Spruce lumber rectangular prisms of 220 by 160 mm were the irradiated wood elements used. The irradiated wooden prisms were cut-outs from wooden ceiling beams. In total, eight prisms of identical profiles were selected. The course of the experiment is described in the table, which also includes the standard deviation or radian. Prior to start of the experiment, each prism was weighed and measured for the appropriate surface and depth moisture content. The initial moisture content varied for the pieces. The different moisture of different elements prepared for the experiments varied from a number of aspects including porosity, proper cracks or volume stability. Inner temperatures of the experimental samples were checked and measured following every single irradiation cycle [1].

3.2 Process of heating and drying

3 cycles of 15 minutes. (15 min of irradiation, 120- minute break), Microwave emitter power: 750-800 W [1].

4 Results and discussion

Following the final irradiation, the elements were left for 18 hours to allow free vaporization of accumulated water vapour. Next, moisture content and weight of the samples were analyzed and the obtained data were compared to initial values.

The entire process is caused by the way how the radiated energy created by a generator is absorbed. The radiated energy is first absorbed by the water molecules closest to the source of radiation. Thus after the first heating cycle the temperature went up only in a certain depth, the rest of the volume is then gradually warmed up by the subsequent cycles [1].

Similar results were then observed in case of the internal temperature then in case of the surface ones, after several cycles of radiation exposition [4]. The values of humidity were recorded at the beginning and also by the end of ongoing measurement process. From these findings it is evident, that not all samples behaved in the exact same manner. Some samples did have a higher moisture content on the obversed side in the final stages, which may be caused by the gradual extrusion of water vapor on the surface shows the initial, mid-time (after stale) and the final weight of exposed elements. As mentioned above, sample No.3 had a higher initial mass because of the rawness of sample.

On the microwave generator is represented together with a funnel waveguide, which were used for the exposition of wooden elements [6].

Table 1. Surface and depth temperatures.

Wood elements		Surface temperatures [°C]				Temperature inside prism [°C]		
		1st heating	2nd heating	3rd heating	Inner temperature	1st heating	2nd heating	3rd heating
Prism no. 1	front	98.9	101.0	106.0	I	27.7	62.9	86.6
	back	27.2	37.1	51.4	II	31.7	74.4	94.7
Prism no.2	front	96.5	106.0	95.1	I	27.0	58.4	95.1
	back	18.1	31.2	53.1	II	30.2	64.8	97.2
Prism no.3	front	84.2	98.0	98.9	I	16.6	37.2	48.5
	back	15.9	20.5	30.6	II	49.8	45.4	58.4
Prism no.4	front	97.2	99.8	105.1	I	26.2	57.1	93.7
	back	25.3	36.2	49.5	II	28.7	61.8	95.2
Prism no.5	front	94.3	97.5	101.7	I	23.3	54.6	91.4
	back	16.4	27.8	49.2	II	25.1	58.5	92.7
Prism no.6	front	95.1	104.6	106.3	I	27.8	59.4	96.3
	back	17.3	29.2	51.1	II	30.6	63.8	98.2
Prism no.7	front	97.8	100.5	106.2	I	26.7	58.2	94.3
	back	26.1	37.4	51.6	II	28.1	61.9	96.2
Prism no.8	front	96.2	105.1	94.9	I	27.1	58.8	95.3
	back	17.6	29.2	51.1	II	30.3	65.1	96.7
Average $\bar{x}$	front	95.0	101.6	101.8	I	25.3	55.8	87.7
	back	20.5	31.1	48.5	II	31.8	62.0	91.2
Median	front	96.4	100.8	103.4	I	26.9	58.3	94.0
	back	17.9	30.2	51.1	II	30.3	62.9	95.7
Dispersion	front	18.6	9.4	21.1	I	12.6	54.2	227.2
	back	20.2	29.0	46.8	II	49.8	57.8	155.8
Standard deviation	front	4.3	3.1	4.6	I	3.5	7.4	15.1
	back	4.5	5.4	6.8	II	7.1	7.6	12.5

 DOI: 10.1051/matecconf/20179301008

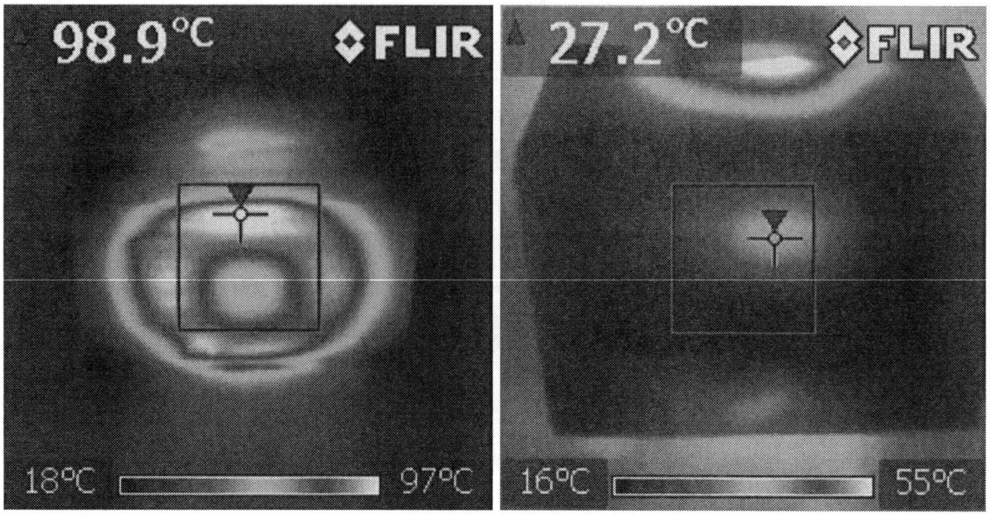

Fig. 3. Temperature field of the prism after the first heating from the back and the front.

Table 2. The moisture and weight values.

Wood elements		Moisture [%]		Weight [g]		
		Initial	120 minutes after heating	Initial	18 hours following heating	Weight loss
Prism no. 1	front	17.1	11.2	5754	5449	305
	back	18.1	14.9			
Prism no.2	front	17.7	19.0	5887	5428	459
	back	18.5	17.6			
Prism no.3	front	18.5	16.9	7955	7318	637
	back	22.0	20.0			
Prism no.4	front	17.5	11.8	5782	5464	318
	back	18.3	15.2			
Prism no.5	front	18.1	19.7	5927	5489	438
	back	18.9	18.3			
Prism no.6	front	17.0	19.1	5836	5417	419
	back	18.3	17.4			
Prism no.7	front	17.2	11.4	5761	5435	326
	back	18.3	15.1			
Prism no.8	front	17.9	19.3	5874	5447	427
	back	18.7	17.5			
Average $\bar{x}$	front	17.6	16.1	6097.0	5680.9	416.1
	back	18.9	17.0			
Median	front	17.6	18.0	5855.0	5448.0	423.0
	back	18.4	17.5			
Dispersion	front	0.2	13.2	496598.0	383315.4	10166.1
	back	1.4	2.8			
Standard deviation	front	0.5	3.6	704.7	619.1	100.8
	back	1.2	1.7			

5 Conclusions

The aim of the experiment was to determine effects of higher initial moisture content of the irradiated material on the penetration of microwave energy and thus on the heating. Test specimens were taken from sawn wood in a so-called "raw" state [6].

As can be seen from the above tables, it has been determined experimentally that it is advantageous for the microwave drying of wood elements to use a lower microwave power and shorter irradiation cycles with longer pauses. This is due to the fact that during long irradiation cycles with higher performance and short breaks between radiation cycles large differences have been analysed between the temperatures on the front and back of the experimental prisms. Following final cooling, shape deformation and creation of additional drying cracks could be observed [1]. With those elements that were irradiated gradually, i.e. using lower microwave power, shorter drying cycles and longer breaks the increase in temperature was gradual, evaporation of water continued in a balanced manner and basically, any deformation of the wood element has been avoided.

In conclusion, the current results of experiments have found that cyclic microwave elimination of moisture in wood elements is effective. It is essential and highly effective to apply cyclic irradiation in the elimination of moisture in wood elements than the same period of time without a cyclic repetition. It is possible to conclude that irradiation using emitters with lower microwave power is fully effective and thus absolutely sufficient. Given the results of experimental work, significant and rapid elimination of weight and moisture has been achieved as compared with the initial and final values. The method of cyclic microwave drying of wood elements appears to be one of the most effective, fastest and thus the most economical methods for local and targeted reduction of moisture content in wood elements and building structures [1].

This paper has been worked out under the project No. LO1408 "AdMaS UP - Advanced Materials, Structures and Technologies", supported by Ministry of Education, Youth and Sports under the „National Sustainability Programme I".

References

1. M. Novotny, K. Suhajda, J. Sobotka, J. Gintar, E. Suhajdova, M. Matl, Wood Res-Slovakia **59**, 389-400 (2014)
2. D. Beckovsky, L. Vackova, T. Beckovska, J. Sobotka, J. Pencik, M. Lavicky, Wood Res-Slovakia **61**, 627-636 (2016)
3. R. Smolka, T. Petricek, P. Kacalek, *enviBUILD 2014 - Buildings and Environment* (Brno, 2014)
4. M. Novotny, K. Suhajda, J. Sobotka, J. Gintar, AMR **1041**, 297-302 (2014)
5. *ČSN P 73 0610, Waterproofing of buildings* (Czech Office for Normalization, Praha, 2000)
6. M. Novotny, J. Skramlík, K. Suhajda, J. Sobotka, AMR **688,** 27-36 (2013)

MATEC Web of Conferences 93, 01009 (2017) DOI: 10.1051/matecconf/20179301009
Building Defects 2016

Design proposal for door thresholds

Radim Smolka[1,*], *Jindřich* Sobotka[1], and *Jindřiška* Šulistová[2]

[1] Brno University of Technology, Faculty of Civil Engineering,
Institute of Building Structures, Veveří 95, 602 00 Brno, Czech Republic
[2] Institute of Technology and Business in České Budějovice, Czech Republic

Abstract. Panels for openings in structures have always been an essential and integral part of buildings. Their importance in terms of a building's functionality was not recognised. However, the general view on this issue has changed from focusing on big planar segments and critical details to sub-elements of these structures. This does not only focus on the forms of connecting joints but also on the supporting systems that keep the panels in the right position and ensure they function properly. One of the most strained segments is the threshold structure, especially the entrance door threshold structure. It is the part where substantial defects in construction occur in terms of waterproofing, as well as in the static, thermal and technical functions thereof. In conventional buildings, this problem is solved by pulling the floor structure under the entrance door structure and subsequently covering it with waterproofing material. This system cannot work effectively over the long term so local defects occur. A proposal is put forward to solve this problem by installing a sub-threshold door coupler made of composite materials. The coupler is designed so that its variability complies with the required parameters for most door structures on the European market.

1 Introduction

Financial incentives, Czech standards and European Union directives is motivating builders to build low-energy and passive houses. As a result, the material characteristics of the products used has improved and continues to do so. At the same time, new construction technologies are being put into practice, and ever increasing demands placed on the economical, technical-physical, technological, hygiene, fire prevention, environmental and aesthetics of buildings. Unfortunately, there are numerous limiting factors which have revealed design and consequently implementation faults in low-energy or passive houses. The biggest issue is the occurrence of thermal bridges and the quality of the implementation of critical details. The lack of prevention of the occurrence of thermal bridges is a frequent problem in the construction of low-energy and passive buildings, particularly in more complicated structure details, which are slightly suppressed in conventional constructions. The requirements for the quality of the design and implementation of a building are therefore increasing, which has a simultaneous impact on costs.

* Corresponding author: smolka.r@fce.vutbr.cz

The demand for disposable packaging is increasing rapidly in the world, with consumption approaching two million tonnes a year in the Czech Republic. Development in this field are closely linked to the use of composite materials including those based on recycled plastics. In the Czech Republic, the quantity of packaging that needs to be recycled is also growing rapidly. Public education and the location of special bins in densely populated areas has contributed to this. Nevertheless, tens of thousands of secondary plastic raw materials, including plastic bottles, end up at domestic waste landfill sites. A polymer becomes a plastic after being mixed and combined with the necessary additives and transformed into a form suitable for further technological processing e.g. grains, powder, tablets, etc. The term "polymer" is understood to mean a chemical substance, whilst "plastic" is a technical material with specific usable properties.

2 Processing method

When choosing the areas for research, consideration was given to the construction details of low-energy and passive constructions. Particular attention was focused on the carcass of a building, which is critical in terms of both design and implementation. As a result, the design of the bottom of a door structure, the plinth of an aired facade and the base area of a roof structure were selected. The most important requirements placed on structure elements are: thermal insulation, airtightness, vapour-tightness, waterproofing, fire resistance, strength and stability, soundproofing, health and safety.

3 Improvement of polymer properties

To improve the polymer properties of products used in construction, antidegradants are added. These additives protect the products on a long-term basis against external influences such as sunlight, atmospheric oxygen, ozone and thermal energy.

Antidegradants can be divided into three categories:

- light stabilizers – absorb UF radiation,
- antioxidants – eliminate oxidation (aging),
- antiozonants – eliminate the influence of ozone.

Other additives that influence the properties of polymeric substances include:

- fillers,
- hardeners – substance strengthening,
- blowing agents – expanded substances,
- pigments – colours of mixtures,
- optically lightening substances – absorption of UV light,
- antistatic additives,
- factices – improvement of surface smoothness,
- adhesive agents,
- agents which decrease combustibility – retarders, extinguishers [1].

These most frequently used additives are normally used in secondary materials – fence blocks, grass blocks, step blocks [2].

4 Structure element of sub-threshold door couplers

The structure details of door panes in constructions have to be determined on the basis of their thermal, mechanical and chemical properties, but also from the point of view of the user. In so doing, it is necessary to eliminate the heat flow below the door frame structure and to connect and terminate the waterproofing and thermal insulation elements.

4.1 Secondary plastic material

This section specifies the secondary plastic raw materials that were assessed and tested.

4.1.1 Selected raw materials and their characteristics

The thermal conductivity coefficient is affected by the basic properties of the individual polymers and by the proportion of additives used to create the plastic material. The value of the thermal conductivity coefficient is reduced by the removal of the plastic filling and the addition of other additives and components (application of a blowing agent, granulate soaking, etc.), as well as by the technology (moulding, injection) used for the production thereof.

Table 1. Measured thermal conductivity coefficient λ [W m^{-1} K^{-1}] values.

Designation	Material	Acronym	Characteristic	λ [W m^{-1}K^{-1}]
1. D)	polyamide	PA		0.196
2. D)	polycarbonate	PC		0.236
3. D. a)	polystyrene	PS	tough	0.345 (max. value)
3. D. b)		PS	brittle	-
4. D. a)	polyvinylchloride	PVC	clear	0.287
4. D. b)		PVC	white	0.315
5. D)	high-density polyethylene	HDPE	black	0.125
1. A)	A1005			0.103 (min. value)
2. A)	A1006			0.152
3. A)	A1007			0.148
1. S)	amorphous blends of styrene acrylonitrile	ABS	red	-
2. S)	amorphous resin	ASA	grey	-
3. S)	Amorphous polycarbonate polymer	PC	clear	0.262
4. S)	polyamide 6	PA 6	brown	0.165
5. S)	semi-crystalline polyethylene terephthalate	PBT	white	0.187
6. S)	amorphous polyetherimide	PEI	brown	-
7. S)	thermoplastic polymers	TPU	mixture	0.152

The results in Table 1 show worse thermal conductivity coefficient parameters compared to the insulation material used today. In terms of the thermal conductivity coefficient the above samples are between three to nine times worse than the thermal insulation normally used [3-5].

4.1.2 Theoretical design of a product for construction

The requirements for products to be used in the construction industry for the elimination of thermal bridges are given. These requirements should be applicable to both new buildings (passive and low-energy constructions) and to existing buildings undergoing reconstruction or modification.

For the purposes of this study structure details for new constructions, particularly for low-energy and passive constructions, have been selected. This selection includes details of corners, horizontal structures, filling of construction structures, masonry footings and the filling of structure openings. The most common threshold arrangement for entrance doors are: aluminium threshold profile with interrupted thermal bridge; barrier-free solution with additional sealing with a movable seal in a special profile embedded in the floor structure.

The proposed sub-threshold door coupler should replace existing substructures in the floor structure. The substructure should consist of at least two boards of recycled plastic and should enable variable connection to the waterproofing structure [6]. To accommodate for varying designs in floor thicknesses and other local influences, the substructure must be a variable arrangement. For such a variable arrangement and for ease of installation, the shape of the substructure boards should be such that the boards fit into or onto each other. They should also be able to be easily cut to size to fit every structure detail (see Figure 1).

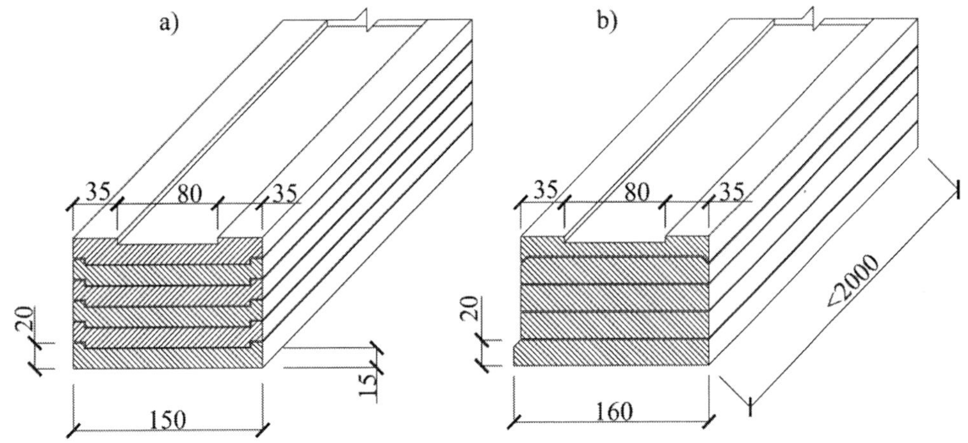

Figure 1. Alternative threshold coupling systems made of secondary raw materials - axonometric view.

The design of the boards and their material characteristics should ensure the distribution of the dynamic load, caused by normal operation, to the subbase layers. The board must also be non-water absorbing and diffusion proof [1, 7-9].

4.1.3 Analysis and comparison of the thermal-technical properties of the structure after theoretical integration of the product

This section illustrates the structure details of a proposed solution which incorporates the proposed product.

For the analysis, the thermal insulant which is normally used in practice is substituted by the proposed sub-threshold door coupler. The door coupler is made of secondary plastic raw materials with a thermal conductivity coefficient of λ=0.103 [$Wm^{-1}K^{-1}$] (the value was taken as the minimum of the measured values).

The thermal-technical behaviour of the recycled plastic insulation block was modelled by means of the ANSYS computing system. For the thermal-technical analysis, the recycled plastic product was embedded into a wall that meets the requirements for passive houses (see Figure 2).

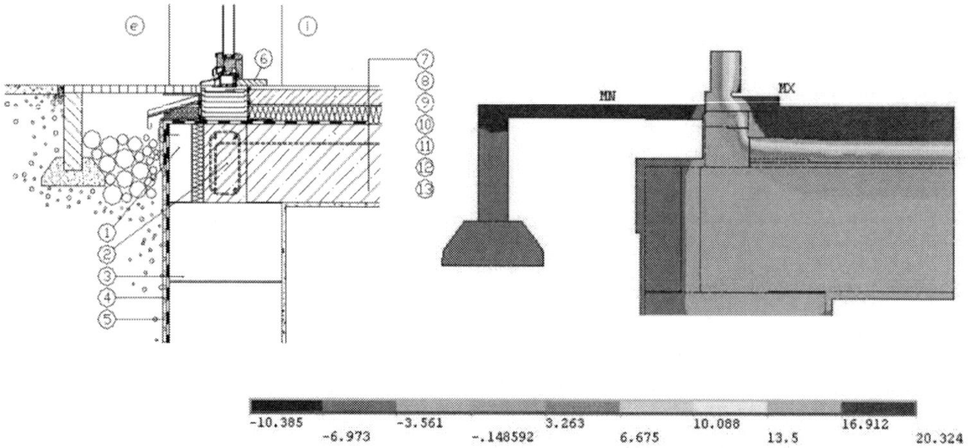

Figure 2. Structural detail of a threshold of a house with a cellar / temperature distribution.

It is clear from the selected and presented example that the sub-threshold door coupler has to be supported by additional thermal insulation. However, this does not have to be under the door structure but can be a part of the thermal insulation of the plinth or subterranean part of the building. Once the value of the thermal conductivity coefficient approximates that of standard thermal insulation (approx. $\lambda = 0.035$ W m^{-1} K^{-1}), the subbase structure could be shifted to the exterior.

5 Conclusion

Materials derived from recycled plastic have great potential for use in a broader range of civil engineering applications. Their technical properties i.e. physical, mechanical and chemical, predispose them for use in structures where conventional thermal insulation is normally used. When using a product derived from secondary raw materials the structure thickness can be reduced. The reason for this is that it is no longer necessary to use other waterproofing or bearing layers because the requirements under technical standards, regulations and laws are all fulfilled at the same time. Another indisputable advantage is the reduction in financial costs during construction. Sub-threshold coupling is only one of the few ways to make use of secondary raw materials in civil engineering [10].

This paper has been worked out under the project No. LO1408 "AdMaS UP - Advanced Materials, Structures and Technologies", supported by Ministry of Education, Youth and Sports under the „National Sustainability Programme I".

References

1. M. Novotný, K. Šuhajda, J. Sobotka, Z. Jiroušek, AMR **1122**, 65-69 (2015)
2. V. Duchacek, *Polymers – production, properties, processing, applications* (Institute of Chemical Technology in Prague, 2006)

3. *CSN 73 0540-2 Thermal protection of buildings - Part 2: Requirements* (Czech office for standards, metrology and testing, Prague, 2011)
4. S.L. Simon, G.B. Mckenna, J. Reinf. Plast. Compos. **18**, 559-571 (1999)
5. L.A. Rorrec–Gomez, R. Yanez-nieto, Thermochim. Acta **140**, 139-144 (1989)
6. J. Plachý, *Bitumen sheet joints under the influence of temperature and time in the composition of single-membrane roofs* (Doctoral thesis, Brno University of Technology, Faculty of Civil Engineering, Brno 2005)
7. L. Matejka, R. Smolka, *Brno University of Technology: Support structures for a doors threshold, EP 2159364 B1*, (patent, Munchen, 2012)
8. J. Pencik, L. Matejka, P. Zitt, *Brno University of Technology: Material deformation testing stand for the long- term monitoring of deformation characteristics under constant pressure, EP 2078948* (patent, Munchen, 2009)
9. T. Petříček, P. Kacálek, R. Smolka, AMR **1041**, 71-74 (2014)
10. R. Smolka, T. Petricek, P. Kacalek, *enviBUILD 2014 - Buildings and Environment* (Brno, 2014)

Contribution to the snow protection solutions for pitched roofs

Jozef Oláh[1] and *Michal* Šida[1,*]

[1] Slovak University of Technology in Bratislava, Faculty of Civil Engineering, Radlinského 11, 810 05 Bratislava, Slovakia

Abstract. This paper presents results of research conducted as part of doctoral dissertation. It deals with pitched roofs defects and failures caused by insufficient snow protection. It also assesses the most common snow retention systems in Slovakia in terms of holding strength and reliability. Selected snow retention systems that are used for different types of roofing were loaded by forces in a direction parallel to the roofing. Paper evaluates all tested snow retention systems and indicates their weaknesses. This paper also deals with friction coefficient between snow and the most commonly used roofing in Slovakia.

1 Introduction

Improper or wrong design of sloped roofs in snow and cold regions, under-dimensioned snow protection systems, leading to the roof damage or endangerment of people or property situated in the potential impact zones of snow or ice, is still reality in Slovakia. There are several reasons for this. The rules resulting from Slovak legislation, according to which the roof snow protection should be designed, is insufficient and unclear, in some cases even contradictory. On the one hand, there is an obligation to use snow retention system for roofs with a slope greater than 25° (Decree of the Ministry of Environment of the Slovak Republic No. 532/2002) [1], on the other hand there is allowed falling snow off the sloped roofs according to standard STN 73 1901 [2] Designing of roofs. Slovak legislation defines no principles of snow retention system design and therefore engineers must rely solely on the recommendations of manufactures. Therefore, this doctoral dissertation was trying to point out pros and cons of the most commonly used types of snow retention systems in Slovakia by measurement of their holding strength and determine recommendations for design of snow retention systems based on findings of measurement of friction between snow and most commonly used roofing for sloped roofs in Slovakia [3-4].

2 Measurement of snow retention systems holding strength

One part of dissertation assignment was to measure holding strength of individual snow retention system elements under laboratory conditions, observe their weak points, evaluate measured results and compare them with values declared by the manufacturer.

* Corresponding author: sidaxmichal@gmail.com

2.1 Method

Three samples of each snow retention system elements were subjected to the laboratory test. Each sample was installed to the test apparatus and loaded by forces in a direction parallel to the roofing. Loading test was cancelled for each element individually according to the test progress. Testing was performed at certified laboratory Microbac Laboratories, Inc., Boulder – Colorado, USA, using electro-mechanical universal testing machine SFM-60. The effort was to select the most commonly used types of snow retention systems in Slovakia. New products, not used, were subjected to the test. A total of five types of snow retention systems were selected for the test.

- **Snow fence type "A"** – linear snow retention system consisting of a special tile, support bracket and snow fence, made of aluminium. This system is designed for tile roofing. Special aluminium tile is hooked over the tiling battens and screwed to them and support bracket is snapped on a tile.
- **Snow fence type "B"** - linear snow retention system consisting of universal support bracket and snow fence, made of galvanized steel with thickness of 4 mm. The system is used for tile roofing. Support bracket is hooked over the tiling battens and screwed to them,
- **Snow rail** - linear snow retention system consisting of support bracket (consisting of two parts) and tubes made of galvanized steel. The system is designed for standing seam metal roof system. Bracket is attached to the standing seam by connecting of both parts over the standing seam and tightening the bolts,
- **Snow strap** – non-linear snow retention system consisting of hooks made of galvanized metal sheet. Snow guards are used for tile roofing and are hooked over the top of tiles,
- **U-shaped snow guard** - non-linear snow retention system consisting of snow U-shaped snow guards made of galvanized metal sheet. The system is used for corrugated metal roofing and snow guards are screwed to the metal roofing using screws with rubber seal.

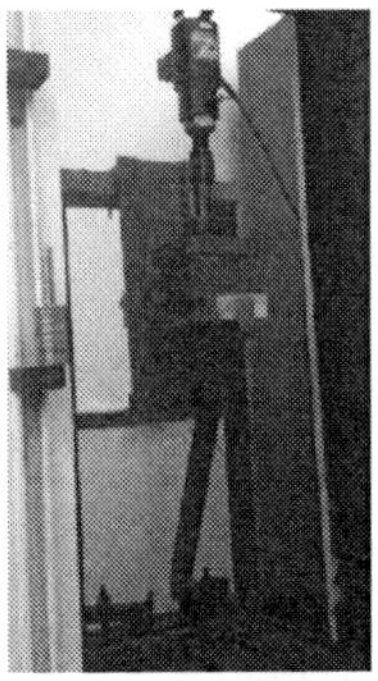

Fig. 1. Snow fence type "A".

Fig. 2. Snow fence type "B".

Fig. 3. Snow rail.

Fig. 4. Snow rail.

Fig. 5. Snow strap.

Fig. 6. Snow strap.

2.2 Results

The most important findings observed during the laboratory testing are following:

2.2.1 Snow fence type "A"

- The average ultimate strength of support bracket snapped to special tile determined by testing was 6,84 kN and exceeded declared ultimate strength 4,0 kN approximately 1,7 times, which corresponds with safety factor specified by Austrian standard Önorm B 3418 – Design and construction of snow protection systems on roofs [5].
- The ultimate strength of the support bracket is given by strength of attachment to the battens, which failed during the testing.

2.2.2. Snow fence type "B"

- Ultimate strength of individual tested support bracket varied widely, which could be caused either by measurement inaccuracies or different mechanical properties of tested brackets. The lowest value of ultimate strength was 1,36 kN.
- The ultimate strength of support bracket is given by quality of particular product – steel grade, quality of processing, storage, handling on construction site etc., since they were plastically deformed.

The manufacturers of both systems, snow fence type "A" and "B", recommend to use them in combination with non-linear snow guards that help to retain a snow blanket over the entire roof surface. Based on this fact and experimental analysis, it can be stated that both tested linear systems are reliable in terms of their holding strength.

2.2.3 Snow rail

- The average maximum strength at which a displacement of support bracket with value of 1 cm along standing seam occurred was 4,63 kN.
- The holding strength of support bracket is given by tightening torque of bolts that are used to attach the support bracket to the standing seam. Of course, recommended tightening torque cannot be exceeded in order to avoid mechanical damage to metal roofing by bolts.
- Calculated allowable load on a single support brackets with span 1200 mm, according to the manufacturer recommendations, is higher than measured holding strength 4,63 kN for roof slopes 0°-23°. The manufacturer probably considered friction between snow and roofing at lower roof angles. For support brackets with span 800 mm is snow retention system design according to the manufacturer recommendations reliable for all roof slopes.

The holding strength of all tested linear snow retention systems depends also on mechanical properties of linear elements (fences, tubes), which were not subjected to the laboratory testing.

2.2.4 Snow strap

- Ultimate strength of individual tested snow straps varied widely, which could be caused either by measurement inaccuracies or different mechanical properties of tested brackets. The lowest value of ultimate strength was 1,07 kN.
- Even value of allowable load on snow strap for the worst scenario (0,83 kN), calculated according to the manufacturer recommendations, was lower than measured ultimate strength (1,07 kN).

- The ultimate strength of snow strap is given by quality of particular product – steel grade, quality of processing, storage, handling on construction site etc., since they were plastically deformed.

This snow retention system, designed according to the manufacturer recommendations in combination with linear types snow retention system, can be stated as reliable.

2.2.5 U-shaped snow guard

- The average measured ultimate strength of U-shaped snow guard was 2,9 kN and exceeded even value of allowable load on snow guard for the worst scenario 0,83 kN, calculated according to the manufacturer recommendations.
- Average strength 0,53 kN, at which start of fastener screws rotation could have been sighted with the naked eye, was lower than value than allowable load on snow guard for the worst scenario. Fastener screws rotation causes that rubber seal does not rest on the roofing and screws penetration is not sealed effectively.
- The holding strength of U-shaped snow guard is given by mechanical properties of roofing since snow guard is fastened to the roofing only.

If roof reliability in terms of watertightness is neglected, U-shaped snow guard designed according to the manufacturer recommendations is reliable in terms of holding strength. However, the roof watertightness is endangered at higher allowable loads, therefore this snow retention fastening system should be stated as inappropriate.

3 Measurement of friction between snow and roofing

Second part of dissertation assignment was to determine static and dynamic frictional coefficient between snow and different types of roofing, in order to determine boundary conditions for the start of snow mass sliding upon individual types of roofing and potential impact zones of snow and ice sliding off the roofs.

3.1 Method

Friction was measured by method, which principle is to measure how large inclination angle between horizontal plane and inclined plane is necessary for a snow sample to start sliding downwards the roofing and for uniform motion of snow sample upon roofing. Static frictional coefficient f_s and dynamic frictional coefficient f_d is given by the equation:

$$f_s = \tan\alpha \,,\; f_d = \tan\beta \,,\; [\text{-}] \qquad (1)$$

For measurement of friction, testing apparatus was designed and constructed. It is a steel frame with adjustable part, with dimensions 1 × 2,5 m, for adjusting the inclination angle in a range of 5 to 80 degrees. Adjustable part is designed in order to change segments with individual types of roofing. During the testing, the ambient and roofing surface temperature was monitored by thermometer Greisinger GMH 3350. The effort was to select the most commonly used types of roofing for sloped roofs in Slovakia. New products, not used, were subjected to the test. A total of four types of roofing systems were selected for the test:

- **Natural clay roof tile,**
- **Galvanized metal sheet,**
- **Asphalt shingle,**
- **Stone coated steel metal roofing.**

3.2 Results

The most important findings observed during the testing are following:

- At negative roofing surface temperatures, between snow and roofing could be created such a strong adhesive forces, that snow does not start sliding upon the roofing even at steep roof slopes.
- Friction between snow and roofing is the lowest at positive roofing surface temperatures.
- Frictional coefficient between snow and roofing is affected by water film created by snow melting, which lubricate roofing surface.
- Frictional coefficient between snow and galvanized metal roofing is affected by water film created by snow melting, which lubricate roofing surface. At high value of snow water equivalent, in terms of tribology, the lubricated friction occurs between snow and roofing if thickness of water film is greater than surface roughness of metal roofing.
- Frictional coefficient between snow and natural clay roof tile is affected by water absorbency of tile, which does not allow the creation of significant water film on its surface.
- Frictional coefficient between snow and asphalt shingle and stone coated steel metal roofing is affected by water film created by snow melting and surface roughness, which reduces the effect of lubrication by melting water. Water film thickness does not reach the height of surface roughness, and in terms of tribology, solid/boundary or mixed friction occurs between snow and these types of roofing.
- **The static coefficients of friction** between the snow and selected types of roofing at positive surface temperatures of the roofing are calculated as the tangent of the angle at which the start of snow sliding upon the roofing occurred.

Table 1. Measured static coefficients of friction between snow and roofing at positive roofing surface temperature.

Type of roofing	β [°]	f_d [-]
Snow / Galvanized metal sheet	10	0,18
Snow / Natural clay roof tile	30	0,58
Snow / Asphalt shingle	40	0,84
Snow / Stone coated steel metal roofing	55	1,43

- **The dynamic coefficients of friction** between the snow and selected types of roofing at positive surface temperatures of the roofing are calculated as the tangent of the angle at which snow was sliding uniformly.

Table 2. Measured dynamic coefficients of friction between snow and roofing at positive roofing surface temperature.

Type of roofing	α [°]	f_s [-]
Snow / Galvanized metal sheet	12,5±2,5	0,22±0,05
Snow / Natural clay roof tile	32,5±2,5	0,64±0,06
Snow / Asphalt shingle	47,5±2,5	1,10±0,10
Snow / Stone coated steel metal roofing	67,5±2,5	2,45±0,30

Fig. 7. Tested types of roofing.

Fig. 8. Testing apparatus.

4 Conclusion

The snow on the roof issue is very broad topic. This issue involves a number of specific problems, which are closely related, resulting from the snow mass on the sloped roofs. One of these specific problems is movement of snow mass upon the roofing and falling off the roof itself. Consequences resulting from the snow and ice falling off the roof could be eliminated by appropriate design of snow retention system. The testing results of all selected snow retention systems confirmed that if you follow the design recommendation of manufacturers, the systems work reliably in terms of holding strength. The results of linear snow retention systems showed that their reliability could be improved by using non-linear elements, which help retain the snow mass over the entire roof surface. Manufacturers of tested linear snow retention systems recommend the same approach. If the maintaining of the roof reliability in terms of watertightness when possible extreme snow load occurs, the snow retention system for standing seam metal roofing (Snow rail) is the most reliable. On the other side, U-shaped snow guard failed this test. At particular loads, the fastening screws started to rotate and buckled the corrugated metal roofing and rubber seal did not rest on the roofing. That is the reason, why fastening of snow guards or entire snow retention system into to the roofing only with penetration is not inappropriate. If the penetration of the roofing is necessary, the snow retention system must be fasten to the support structure of the roofing. It is also necessary to create the rules in form of standard or some regulation for design and construction of snow protection system on sloped roofs, in order to prevent chaos in Slovak legislation associated with snow protection design and eliminate improper design and construction resulting in roof failures or catastrophes caused by snow falling off the sloped roofs.

References

1. *Decree of the Ministry of Environment of the Slovak Republic No. 532/2002* (2002)
2. *STN 73 1901:2005, Designing of roofs* (2005)
3. MCA Guidelines for Commercial Roofing, Metal Roof design for Cold Climates
4. P. Kollar, Building research journal, **53**, 137-154, (2005)
5. *ÖNORM B3418:2012 Planung und Ausführung von Schneeshutzsystemen auf Dächern* (2012)

MATEC Web of Conferences **93**, 02001 (2017) DOI: 10.1051/matecconf/20179302001
Building Defects 2016

Polyurethane adhesives in flat roofs

Markéta Bogárová[1,*], *Jindřich* Stodůlka[1], and *Karel* Šuhajda[1]

[1]VUT FAST, Ústav pozemního stavitelství, Veveří 331/95, 602 00 Brno, Czech Republic

Abstract. It is necessary to stabilize individual layers of flat roofs, mainly because of wind suction. Apart from anchoring and surcharge, these layers can be secured by bonding. At present gluing is an indispensable and widely used stabilization method. On our market we can found many types of adhesives, most widely used are based on polyurethane. This paper focuses on problematic about stabilization thermal insulation from expanded polystyrene to vapor barrier from bitumen. One of the main issues is to calculate the exact amount of adhesive, which is required to guarantee the resistance against wind suction. In this problematic we can not find help neither in technical data sheets provided by the manufactures. Some of these data sheets contain at least information about amount of adhesive depending on location in roof plane and building height, but they do not specify the strength of such connection. It was therefore resorted to select several representatives polyurethane adhesives and their subsequent testing on specimens simulating the flat roof segment. The paper described the test methodology and results for two types of polyurethane adhesives.

1 Introduction

Flat roofs are mostly designed as single-shell or double-shell. Single-shell roofs provide all function with just one shell that is formed from a few layers. Every layer in the roof shell has to be secured from wind suction.

Among of all sorts of stabilization, gluing is technology that is the most inclinable to extreme conditions. In this case it means especially planeness, cleanliness of base and weather conditions. On the contrary, big advantage of stabilization by gluing is un-overloading of construction, it doesn't come to perforation of lower layers, neither to bringing point thermal bridges into construction. In terminological formulation, gluing is the layer connecting two near-by layers of the roof shell. This connecting layer is designed of proper adhesives intended for these purposes. It is preferred to use adhesives which are not able of permanent plastic deformation by expected temperatures after application.

2 Polyurethane adhesives

Polyurethane adhesives rank among adhesives based on synthetic base indurative with access to air humidity.

* Corresponding author: bogarova.m@fce.vutbr.cz

Double-component polyurethane adhesives are based on reaction of multi-substantial alcohols (polyols) with multifunctional isocyanates. This reaction proceeds very fast attended by creation of dense polymer web.

Mono-component polyurethane adhesives were made of double-component so that polyol was kept to react with abundance of isocyanate. Therefore there were groups of isocyanate that didn't react with each other. These are sensitive to water, alcohols, amines, acids etc. Adhesives setting happens by activity of air humidity in three steps. In the first step it comes to reaction of isocyanate with water and therefore to creation of unstable carbamic acid. The second step is disintegration of karbam acid to carbonic oxide and amine. And the third step is reaction of amine and isocyanate and creation of polyurethane linkage [1].

In the most cases of gluing in flat roofs in the Czech republic, there are used mono-component polyurethane adhesives.

2.1 Gluing issue

When we use polyurethane adhesives as connecting layer and core method of stabilization of flat roof shell as well, many questions about function of glued system arise. Mainly it is about amount of used adhesive and accompanying stability of glued joint. Of course, there are other influences in this whole process such as material of connecting layers, extreme conditions or type of adhesive.

Composing of glued joints in the flat roof hasn't standards or legislatively formed any parameters, if it is about amount of glue or stability of formed joint. Technical data sheet from manufacturers often contain advised amount of adhesive, but no reached stability of glued joint.

By this reason, it came up to testing of glued joints. Testing methodology was created that was based upon ČSN EN 1607 Thermo-insulating products for building applications – Determination of tensile strength perpendicular do faces [2-5].

2.2 Testing methodology

In this methodology, it was necessary to design test model and test apparatus that would be easy and repeatedly attached to test model and also vised into press. In this article, one type of test model will be described with using two types of polyurethane adhesives.

2.2.1 Test model

Test model consists of modified asphalt strip and expanded polystyrene EPS 150 S. These two layers were connected by polyurethane adhesive or the testing joint. To create this test model, it was necessary to provide these two layers with desk tough enough to transfer tension power of whole surface from test apparatus to test joint. Therefore, beech plywood was chosen for its mechanical features. Connection with tough desks was solved with melting of the asphalt strip on its whole surface and also with gluing to polystyrene with polyurethane adhesive on its whole surface.

Fig. 1. Creation of test model with testing joint between asphalt strip and thermo-isolation.

2.2.2 Test apparatus

As you can see on Fig. 2, test apparatus was made of iron plate with 8 holes for mechanical attaching to test model with screws and with hard-faced unwrought extension hexagonal female screw with thread. Others parts of test apparatus were threaded rods used for attaching to jaws of the press.

Test apparatus was always put to the center of test model and thus axial stress was secured. Used press has fixed upper jaws and the speed of stalling of lower jaws was set to 10 mm per minute. Record of grow of tension stress, time and moving of crossbar till to failure of test model was taken in this measurement. Test result was the maximum tension stress above all.

Fig. 2. Iron plate of the test apparatus on the left and gripping of the test model in press on the right.

2.3 Experimental measurement

The experimental part of this article is about comparison of two types of polyurethane adhesives by using sample size 250×250 mm with polyurethane adhesive in two stripes. Gluing itself happened in outdoor condition with 21°C temperature and 88% relative air humidity. Following testing was taken after two weeks in lab conditions with 23±2°C air temperature. There was used 10 pcs of test samples per each testing adhesive´s type, still in some cases distortion outside of tested joint happened.

2.3.1 Test model with tested adhesive number 1

Table 1. Results of experimental measurements.

Test sample nr.	1	2	3	4	5	6	9
F_{max} [N]	2846,9	3865,7	4015,8	3633,1	3267,0	3318,4	2813,1
σ_{mt} [kPa]	45,6	61,9	64,3	58,1	52,3	53,1	45,0
$\bar{x}$ σ_{mt} [kPa]	54,31						
σ [kPa]	6,96						
V_x [%]	12,81						

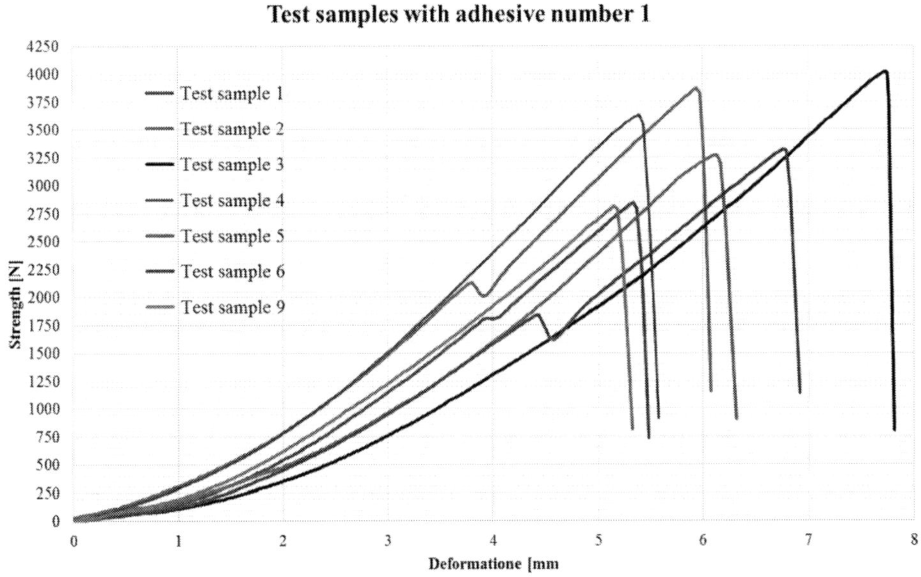

Fig. 3. Graph of results of individual test samples.

2.3.2 Test model with tested adhesive number 2

Table 2. Results of experimental measurements.

Test sample nr.	2	3	4	5	7	9	10
F_{max} [N]	2183,7	2762,0	5128,0	2313,8	1575,2	3821,8	5162,1
σ_{mt} [kPa]	34,9	44,2	82,1	37,0	25,2	61,2	82,6
$\bar{x}$ σ_{mt} [kPa]	52,45						
σ [kPa]	21,43						
V_x [%]	40,86						

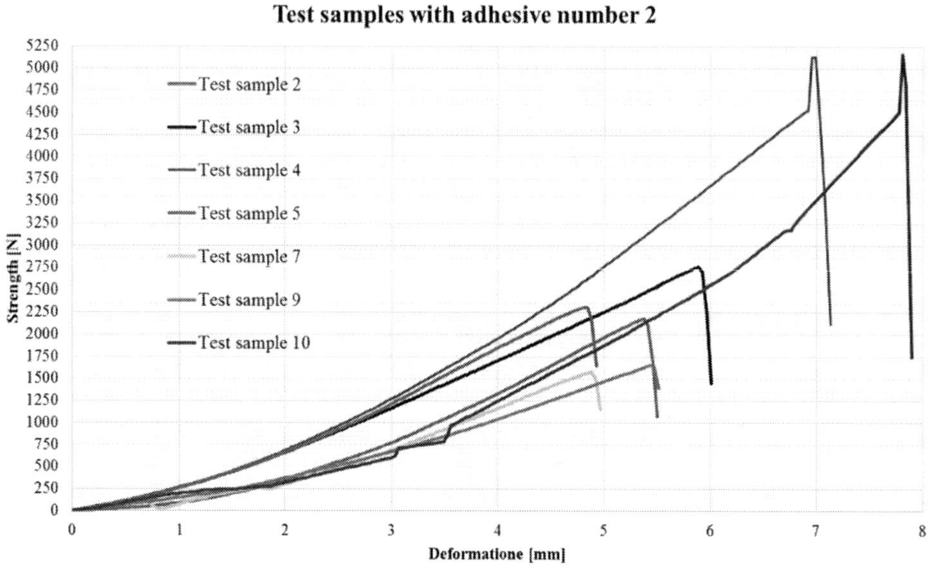

Fig. 4. Graph of results of individual test samples.

2.3.3 Evaluation

In this article above, there are presented results for two polyurethane adhesives using the same composition of test model. As you can see on Fig. 3, the adhesive numbered 1 embodied average value of tension 54,31 kPa by directive deviation 6,96 kPa. Adhesive numbered 2 embodied similar average value of tension as adhesive numbered 1 and it was 52,45 kPa, but with directive deviation 21,43 kPa. From the graph of adhesive nr. 2 we can see that test samples 4 and 10 embodied markedly higher values and with distortion it has come to detachment of considerable part of polystyrene. In view of the fact that all samples were executed by the same extreme conditions and by using of the same materials, we can ascribe this effect to just varied properties of concrete polyurethane adhesive in one pack or to heterogeneity of the structure of expanded polystyrene. Of course, in this whole process we must apply the human factor, but parameters of particular layers are always checked (e.g. weight of adhesive, position of gluing stripes etc.) Considering this we can assume deviations created by executing test apparatuses cannot have large effect upon results. Also in others tests samples, it came to adhesive breakage between adhesive and polystyrene. In some test samples, adhesive embodies better connectivity with polystyrene than with asphalt´s surface. This fact has been demonstrated by dividing adhesive between surfaces.

3 Conclusion

There were stated only results of measurements within the scope of two types of test models in this report. But also from these result we can see that gluing embodies strength values in such a dispersion. This fact is affected mainly by changing surface of concrete adhesive, but as it came from other measurements it's affected by many extreme conditions/side effects.

This article was supported by specific reserch project: FAST-S-16-3345 "Verifikace chování a rozvoj konstrukcí a konstrukčních prvků na bázi dřeva a kombinovaných konstrukcí ze dřeva a železobetonu" on Faculty of Civil Engineering VUT in Brno.

References

1. Basic theory of gluing. ABClepidla.cz, (2013), online: http://www.abclepidla.cz/Files/file/files/Teorie_lepeni.pdf
2. *CSN EN 1607. Thermal insulation products for building applications – Determination of tensile strength perpendicular to faces* (ÚNMZ, Prague, 2013)
3. *CSN 73 1901. Designing of roofs – Basic provisions* (ÚNMZ, Prague, 2011)
4. J. Plachy, J. Musílek, Material for building, **19** (3), 33-37 (Business Media CZ s.r.o., Prague, 2013)
5. A. Fajkos, M. Novotny, *Roofs – basic constructions* (Grada Publishing, a.s., Prague, 2003)

Defects and behaviour of inverted flat roof from the point of building physics

Ivan Misar[1*] and *Marek* Novotný[2]

[1]A.W.A.L. s.r.o., Eliášova 20, 160 00 Prague 6, Czech Republic
[2]Faculty of Architecture, Czech Technical University, Thákurova 9, 169 00 Prague 6, Czech Republic

Abstract. One of the most discussed flat roof structures during the last 20 years is a structure called inverted roof, where the main thermal insulation layer is placed above the main waterproofing system. The reasons why this type of flat roof is or could be chosen are more less clear. Usually it is the intention to protect the main waterproofing system, usually of synthetic or bituminous membranes, against the impact of outdoor air thermal changes, against any prospective mechanical damages and also to reduce risk of water vapor condensation in the structure. This type of structure could help to solve the vapor/thermal difficulties during the design of the flat roof over the space with higher indoor air humidity like swimming pools or specific industrial processes. Due to the higher rate of safety against mechanical damage it is also used quite often in the case of the design of the roof terraces or roof gardens. Nevertheless, the correct attitude during the design of the structure is to take into considerations all possible aspects including the defects and problems which are most typical for each one type of structure. This paper is willing to give the brief overview of the typical defects for inverted flat roofs and to contribute a little to the understanding of commonly discussed effect of undergoing water beneath the thermal insulation itself and decreasing thus the thermal protection efficiency as well as the inner surface temperature.

1 Introduction

The flat roof is one of structures highly affected with the external climatic conditions. Nevertheless, this structure is highly important for the durability and longevity of the building structure as a whole. One specific solution of the structure design is considered to be modern but still provoking a lot of questions. This is the solution based or related to inverted flat roofs. Composition of such structure is known. It is based on the layer of waterproofing liner, or monolithic – in-situ created from liquid applied waterproofing, or prefabricated flexible sheets, and on the layer of thermal insulation placed above the waterproofing itself. The variety is of the liner types is high. The choice of the thermal insulation is limited by the request of low water absorption. At the moment only one product could be taken as a suitable choice for thermal insulation it is extruded polystyrene board [1-8].

* Corresponding author: misar@awal.cz

The crucial question which was highly discussed in the past decade is the heat loss by cold rain water going through the joints between the boards of thermal insulation itself and going towards the water outlet system beneath this insulation, respectively directly on the waterproofing. This phenomenon could even result into the decreasing of internal surface temperature below the dew point temperature and cause thus the vapour condensation on this surface. The most serious period is the snow melting period.

The public opinion presumes the main volume of rain water is going above the insulation boards. The ratio of the water undergoing the thermal insulation boards is not well defined, still.

In order to help us with the orientation in this problematic there was an experiment provided.

2 Description of Experiment

2.1 Definition of Main Task

The experiment itself was supposed to help within description and definition of the ratio of rain water undergoing the thermal insulation. Co-relation with the simulated flat roof slope, upper protective layer, either gravel ballast or paving on the sand or gravel based stabilization layer and the presence of so called water barrier was investigated.

2.2 Consequential Tasks

During the experiment itself a few of next questions or tasks occurred:
- to define the co-relation between the type of separation textile installed on the insulation boards,
- to describe the possibility of tightening of the extruded polystyrene board joints,
- to check the influence of joints and joint imperfections to the heat transmission of base structure by the thermo-vision inspection,
- to define the reducing coefficients (factors) of the **ΔU**-value, for inverted roofs according to the slope, outlet system and rain water barrier installation on the upper thermo insulation boards surface or on any other water tightening system of joints.

2.3 Construction of Testing Equipment

The wooden frame and OSB structure was used to prepare two testing boards of 1x2 m of dimensions. One should simulate the behaviour of flat roof drained by the inner gully and the second one the drainage ensured through the linear flashing on the edge of the roof.

The waterproofing was done from flexible prefabricated SBS modified bituminous membranes with the slate flakes self protection. The thermo insulation boards of extruded polystyrene of 100 mm of thickness were then placed into these frames. Perimeter joints were sealed with the silicone mastic in order to reach the same ratio of joints as in the reality. The AL gutter flashings were installed in the groove 1 cm under the upper edge of the XPS board to collect the rain, resp. artificial rain, water going above the XPS itself.

The composition above the XPS boards was variable. Gravel ballast screed on the non-absorbing separation textile 200 g/m^2 or the ceramic tiles on sand screed on the same textile were installed. Consequently, the gravel ballast screed on the breathing membrane, called lately as a water barrier, was tested as well. The overlaps of this breathing membrane were loosely laid with the 20 cm of width or tighten by the adhesive tape. The next variant was to tighten the XPS boards joints with the compressed taped installed into the gaps.

 DOI: 10.1051/matecconf/20179302002

Fig. 1. Water collection system and structure – frame A (drainage through flashing simulation).

Fig. 2. Water collection system and structure – frame B (drainage through inner gully simulation).

DOI: 10.1051/matecconf/20179302002

3 Results

The ratio of undergoing water in the various roof slopes in condition of artificial rain was measured. The ratio of undergoing water was on the contrary with the public opinion quite high and that it decreased within the increasing slope, depending also on the scheme of water outlet. The undergoing water flow is not steady but it concentrates around the joints of XPS boards, mainly around of those with improper installation. Thermo-vision photos are one of the evidence. It seems that approx. 50% of undergoing water flow rather through joints. It is clear that any mathematical model would be, more less, theoretical and not fully corresponding with the reality.

The quite surprising result was reached in case of standard breathing membrane with the low s_d – equivalent diffusive thickness, under 0.1m, placed in the position of separation layer between the XPS insulation boards and the gravel ballast layer. The results became to be nearly inverted. It seems it would be very efficient for the future to implement this result, means breather membrane installation instead of standard geo-textile one, which is quite commonly used nowadays, as separation layer, to the practice to decrease the heat loss and risk of vapour condensation on internal surface by the undergoing water. This breather membrane takes the function of certain rain water barrier in the structure. In case of indoor high air humidity, it would also slightly reduce the risk of accidental vapour condensation on the inner roof structure surface.

The moisture content by weight up to 4% was measured in the middle of XPS board, after 2.5 years of exposition and beneath the geo-textile of 300 g/m^2.

The measurements in the joint area were much more interesting. In these cases, the increased level of moisture was found, mostly in the layer under horizontal surface of semi lock and in the close location.

This fact can be taken as one of evidences of diffusive absorption existence in this part. Consequently, the reduction of thermal insulation efficiency with increased heat conduction coefficient by the moisture can be expected here as well.

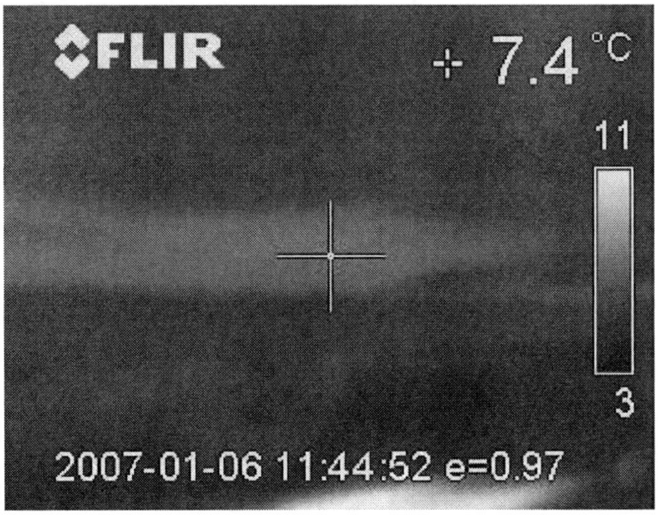

Fig. 3. Approximative surface temperature measurement at the imperfection of the joint location on frame A.

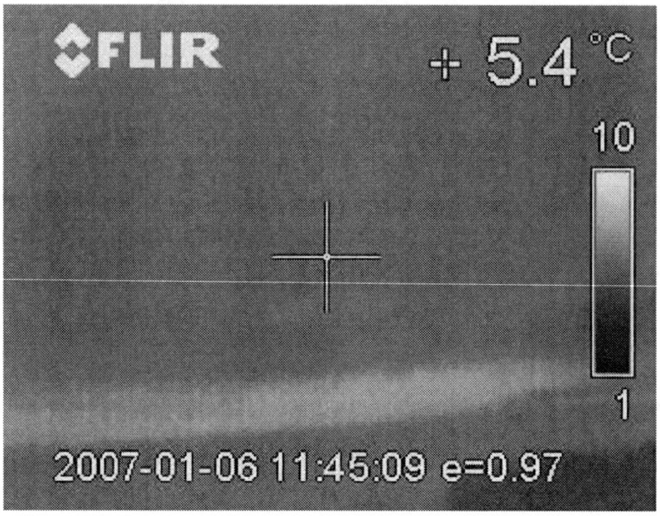

Fig. 4. Approximative surface temperature measurement at the main area of frame A.

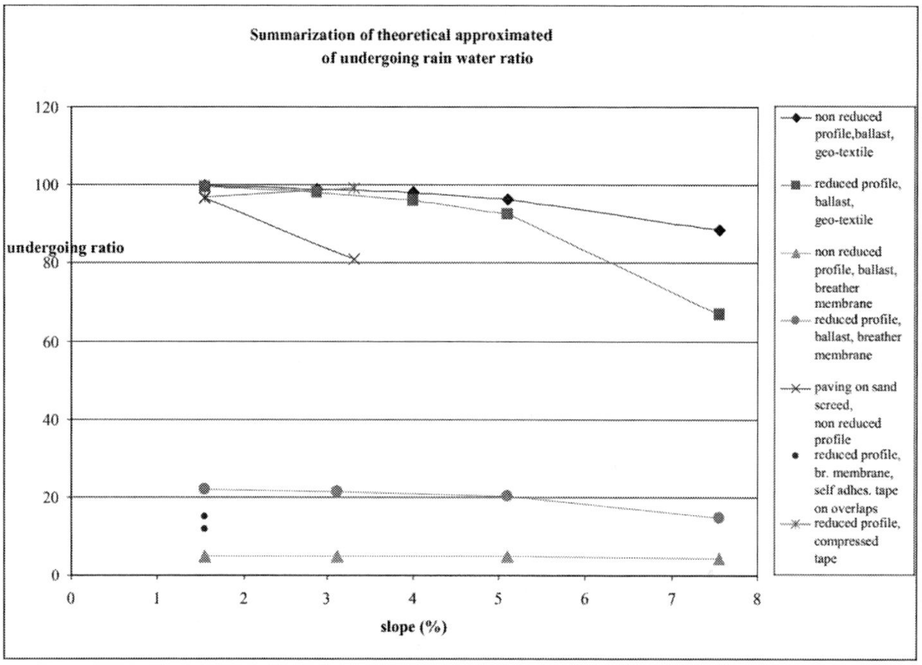

Fig. 5. Summarization of theoretical approximated functions of undergoing rain water ratio in relation with slope.

If the breather membrane will be installed instead of the standard separation geotextile we can for the results representation of the influence use the factor **fw**,

for non-reduced flow profile (sample A):

$$fw = 0.05 \tag{1}$$

for reduced flow profile (simulation of inner gully (sample B):

$$fw = 0.22 \quad (2)$$

Slope reduction coefficient could be represented by slope factor **fs,** expressing the ratio of undergoing water beneath the thermal insulation boards to the all rain water falling down to the roof surface in relation with slope.

for non-reduced flow profile:

$$fs = 1\text{-}0.038 \cdot \sinh(s^2/31) \quad (3)$$

for reduced flow profile (simulation of inner gully):

$$fs = 1\text{-}0.05 \cdot \sinh(s^2/22) \quad (4)$$

where s is a slope of roof structure in %.

These factors could be implemented as reduction coefficients for **ΔU-value**. The equation involving the slope and rain water barrier for **ΔU-value** for inverted roofs would be:

$$\Delta U = 0{,}04 \cdot fw \cdot fs \cdot p \cdot (R_1/R)^2 \; [\mathrm{W/(m^2 \cdot K)}] \quad (5)$$

where ***p*** is average rain water amount in mm/day during the heating season,
R_1 is thermal resistance of XPS layer above the waterproofing system,
R is common thermal resistance of structure in $(m^2 \cdot K)/W$
Fw is rain water barrier installation factor (breather membrane),
fs is slope factor.

4 Conclusions

The experiment has proved the rain water undergoing thermal insulation boards of inverted roofs is playing the significant role within the heat loss through the inverted roof structure which should be a subject of our interest for the future.

On the base of above described experiment we can realize that the influence of water barrier, made of breather membrane, will reduce the ratio of undergoing rain water many times. No significant reduction of the undergoing water ratio was measured when the seams were glued together or when the thermal insulation boards were sealed by compressed tape.

The usage of rain water barrier shall be recommended for the practice. We can still see a wide field of the application physics works for next experiments (e.g. determination of influence of breather membrane quality to the described effect).

On the base of the experiment it is able to say that the drainage of rain water in case of inverted roofs should be preferred through inner gully.

We can say on the base of co-relation between the undergoing water ratio and the type of water drainage system that the reduction of the space between the under surface of insulation boards and the waterproofing plays the role. We can dare to say we shouldn't keep this contact space more permeable for the water. Thus I can dare to suggest avoid the installation of any studded foil under the insulation boards.

References

1. Ch. Curtoise & coll, *Practical Guide to Flat Roofing; Pitsburg Corning Europe N.V.* (1999)
2. A. F. Nielsen, E. M. Paulsen, *Thermal design of highly insulated inverted roofs,* (Norvegian Building Research Institute, Trondheim, Norway)
3. H.M. Kunzel, K. Kiessl, *Assessing Precipitation Heat Losses of Protected Membrane Roofs* (Hemtie Closing Seminar, Finland 1996)
4. J.C. Beech, G.K. Saunders, *The performance of Lightweight Inverted Flat Roofs*, (Building Research Establishment, Princes Risborough Laboratory Buckinghamshire, UK)
5. H.M. Kunzel, K. Kiessl, *Moisture Behaviour of Protected Membrane Roofs with Greenery*, (Frauhofer-Institute of Building Physics)
6. K. Halahyja, B. Beťko, K. Bloudek, P. Tomašovič, P., J. Puškáš, *Stavebná tepelná technika, akustika a osvetlenie technika* (SNTL, 1985)
7. M. Novotný, I. Misar, *Ploché střechy* (Grada, Praha 2003)
8. V. Vymětalík, *Tepelně izolační vlastnosti inverzních střech (Thermoinsulation Abilities of Inverted Roofs)*, (ČVUT, Prague, 2007)

Analysis of roof membranes damaged by mechanical and climatic loads – pilot research

Jakub Čurpek[1,*] and *Jozef* Oláh[1]

[1]Department of Building Structures, Faculty of Civil Engineering STU in Bratislava, Radlinského 11, 810 05 Bratislava, Slovakia

Abstract. Realization of roof construction has many hidden risks, especially in buildings with diverse architecture. There is a problem about cooperation of individual works (vertical and horizontal constructions) on roof in this type of architecture, during the process of realization. Purpose of this research is to reveal risks in form of the group of major source of defects by mechanical damages. The most often types of mechanical damages were chosen in this research, which then were applied on individual types of roof membranes. Response of this damage was found out during the test procedure of water pressure by special laboratory machines. Furthermore, samples of roof membranes were subjected to the Impact test, which was actually focused on damage by hailstone impact from the atmosphere. The final outcomes of the measurements show that the material composition of each roof membrane can influence their whole waterproofing after application of certain type of mechanical damage. In the Impact test, samples were suffered from impact of the hails. This test signified that the choice of base material of thermal insulation below the roof membrane plays an important role.

1 Introduction

Fundamental property of roof is the ability to resist negative atmospheric conditions during its whole lifetime. However, certain potential risks can cause damage of waterproof membrane within realization of a building. The dangerousness is hidden in the craftsmen's cooperation between to each other on a roof and adjacent construction of building. In the all phases of project it is necessary to have in mind that except of traditional designed loads (people, snow, etc.) also human behaviour factor is there. This special load is unpredictable and it cannot be specified in design process. Unfortunately, many workers on a site do not have only a marginal knowledge about function and manipulation of roof membranes and usually could inconsiderately damage them [1]. In doing so, one little damage on the membrane can cause really huge problems on the whole new building in the nearest future. Mostly, foremen try to obscure these problems in the effort to reach work time schedule and get a reward. In the many cases, building could not be able to undergo final inspection due to occurrence of waterproof failures. In addition to, special climatic conditions as hailstorm impact could be able to cause huge damage of membrane placed on some certain

* Corresponding author: jakub.curpek@gmail.com

type of thermal insulation material. Physical properties, especially density and elasticity of underlay material can influence response of membrane under immediate impact of hails. The effort is to find the right way of explaining this influence between damages and whole waterproof functionality of roof.

2 General example of damage

Especially, it is requisite to give better attention during the realization of architecturally shaped objects where generally flat roofs prevail. This type of architecture with gradually changed floors generates a lot of problems with cooperation each step of work on the vertical and horizontal parts of building construction (roof/facade). In many cases, flat roof could provide good place for storing building material, site equipment or scaffold, Fig. 1c. General example of this type of building could provide recent restoration of roof on the building of Slovak University of Technology in Bratislava, Fig. 1a. Many damages of waterproof membranes have been done due to unprofessional and irrational working operations, Fig. 1b. These recent circumstances in practise make up the main reason why this pilot research has been done.

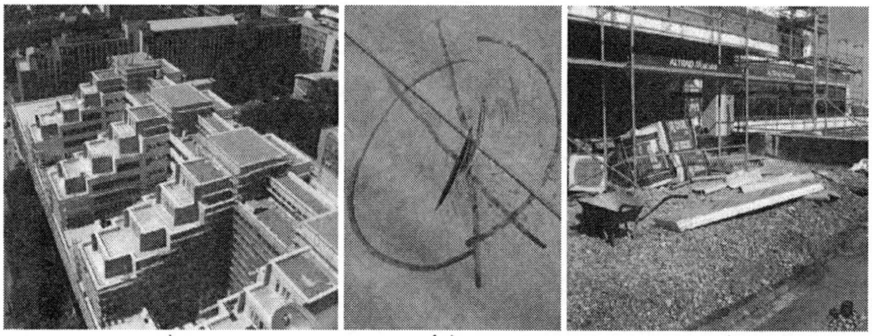

a.) b.) c.)

Fig. 1. The scene of roof restoration: a.) Faculty building of STU in Bratislava, b.) Mechanical damage PVC-P waterproof membrane, c.) Realization of horizontal and vertical construction of STU.

3 Aims and methodology of the analysis

Waterproof properties of roof membranes are determined mainly by each component part used in the whole material structure. The main goal of this research is investigation of dependency between material compositions of the foremost using membranes and mechanical and climatic damage. Analysis of this dependency function was found out by experimental measurements realized in the special laboratories. Two types of analysis have been done in this research: analysis of mechanical damage on membranes caused by human factor and its influence on membrane waterproofing, analysis of hailstorm impact on membranes subjected under laboratory temperature 20 °C placed on different base material. Novel types of samples without any initial damages were used in the experimental measurements as well as samples taken from a site. Consequently, types of mechanical damage were chosen and implicated on membrane samples: 50 mm long knife cut (full and half of thick sample), nail perforation, screw perforation, Fig. 2.

Material of each used membrane samples was chosen as the most used material in a practise: modified asphalt styrene butadiene styrene (SBS) sheets (with fined-grained and coarse-grained strewing), plasticize polyvinyl chloride PVC-P foils (unreinforced, reinforced by polyester lining), ethylene propylene diene monomer (EPDM) foils (unreinforced, reinforced by polyester lining), thermoplastic polyolefin (TPO) foils.

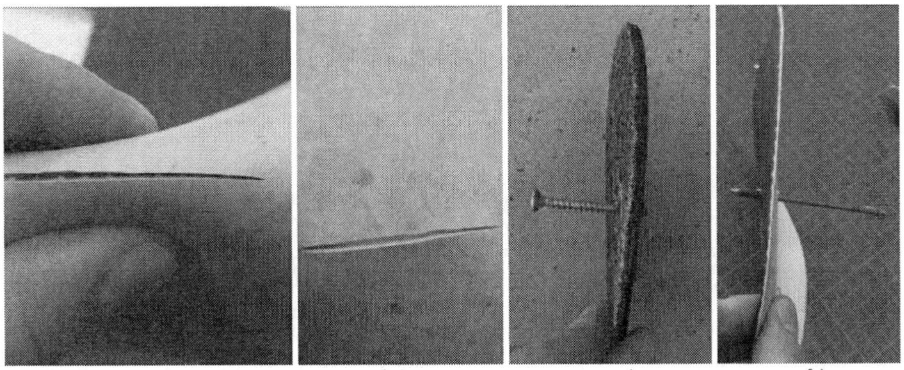

a.) b.) c.) d.)

Fig. 2. Types of mechanical damage: a.) Half-cut knife, b.) Full-cut knife, c.) Nail perforation, d.) Screw perforation.

3.1 Low water pressure testing procedure

Tests were carried out by using EN 1928 method A to determinate watertight properties of investigated samples [2]. Procedure triple tester, Fig. 3a, provides measuring three samples at the same time under maximum water pressure 12 kPa during 24 hour test period. Each sample was cut out for circle shape (ϕ 200 mm) and then place upper side downwards into chamber, covered by filtrated paper and glass plate with a flange ring, Fig. 3b. Tightness of chamber is ensured by three wing screws situated around the ring. Filtration paper could provide instantaneously identification of water penetration through membrane. After checking the whole corrections of apparatus, opening of inlet valve provided flowing of coloured water into chamber and subsequently started to press on the samples until water column reached maximum value 1200 mm. Samples were observed under the lower pressure during water pumping. If certain sample withstood process of filling up the water column then sample was leaved for observation during 24 hour.

a.) b.)

Fig. 3. Laboratory equipment for low-pressure water testing procedure: a.) Watertightness Tester WSP 3600, b.) Putting samples into the chamber.

3.2 High water pressure testing procedure

This procedure could provide testing only one sample by one-time period in compliance with EN 1928 method B [2]. Samples were cut out circle shape (ϕ 130 mm) and then place upper side downwards into apparatus and covered by a circular steel disk containing four slots, Fig. 4b. Cone-shaped covering was put on the disk and tightened by a clamp. Thereafter, opened inlet valve provided water flowing into the apparatus that immediately caused water pressure on the sample, Fig. 4a. Value of water pressure could be adjusted by throttle valve to maximum value 60 kPa.

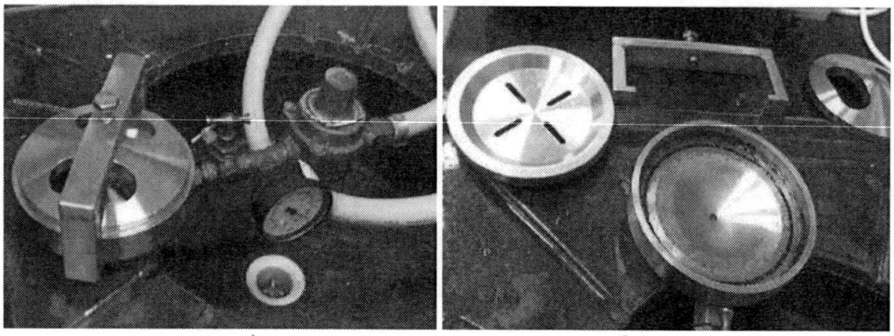

a.) b.)

Fig. 4. Laboratory equipment for high-pressure water testing procedure: a.) Apparatus in process, b.) Each parts of apparatus.

3.3 Impact test

The principle of this test is about simulated impact of a certain weight on the membranes [3] with different underlay material: expanded polystyrene (EPS), mineral wool (MW), polyisocyanurate foam (PIR) and steel plate with 3 mm of thickness. This weight is represented by cylindrical steel drop mass (500 g) with a fixed puncturing spherical-shaped tool (ϕ 12.7 mm), whole weight is attached on a release mechanism, Fig. 5b. The drop height was set up 2 m from the bottom of the puncturing tool to the upwards surface of the membrane. Each sample was burdened by ballast steel ring (2 kg) with inner diameter of 100 mm, Fig. 5c. Investigation was not only focused on the membrane damage but also underlay material damage by weight impact. If it could not be possible visually to determinate character of damage, then sample was subsequently subjected under water pressure testing procedure.

a.) b.) c.)

Fig. 5. Impact test equipment: a.) Whole apparatus, b.) Trigger with a setting device for variable drop, c.) Ballast ring on the sample.

4 Results of measurements

4.1 Results of water pressure testing procedure

Every type of mechanical damage could influence waterproof functionality of membranes in a different point of view. Not all damages were applied on the each sample as well as not all samples were subjected under all testing procedures during this pilot research. The biggest distinctions can be observed on the PVC-P and EPDM foils which were tested as reinforced and unreinforced. Polyester lining could significantly influence waterproof function of membranes for the sake of it could work like conductor and provide transport of water to other side of a membrane. EPDM foils have great elastic properties so they could seal themselves around a puncturing tool. Both types of PVC-P foils damaged by 50 mm half knife cut could be able to withstand 24 hours under low water pressure. Half-cut damage cannot remarkably influence waterproof properties of membranes. Asphalt sheets were able to enclosure perforating tool by asphalt matter and resist under water pressure. Additionally, surface grain structure of asphalt sheets plays certain role in testing. However, these outcomes do not represent membrane performance during their whole life-time but only the very narrow time step.

Table 1. Results of low water pressure testing procedure.

Sample	Mechanical damage	Water pressure [kPa]	Evaluation
PVC-P, reinforced	50 mm full cut	1.10	water penetration
PVC-P, reinforced	screw perforation	2.80	water penetration
PVC-P, unreinforced	50 mm full cut	12	withstood 24 hours
EPDM ,unreinforced	50 mm full cut	12	withstood 24 hours
EPDM, unreinforced	nail perforation	2.10	water penetration
EPDM, reinforced	50 mm full cut	5.15	water penetration
TPO	50 mm full cut	5.60	water penetration
TPO	nail perforation	12	water penetration (during 24 hours observation)
TPO	screw perforation	5	water penetration

Table 2. Results of high water pressure testing procedure.

Sample	Mechanical damage	Water pressure [kPa]	Evaluation
asphalt, fined-grained	none	60	water penetration
asphalt, fined-grained	50 mm half cut	60	withstood 24 hours
asphalt, coarse-grained	no one	60	withstood 24 hours
asphalt, coarse-grained	screw perforation	60	withstood 24 hours
asphalt, coarse-grained	nail perforation	60	withstood 24 hours
asphalt, coarse-grained	50 mm half cut	60	withstood 24 hours
EPDM, reinforced	none	45	water penetration
EPDM, unreinforced	none	60	withstood 24 hours
PVC-P, reinforced	none	60	withstood 24 hours

4.2 Results of Impact test

PVC-P samples (reinforced – taken directly from the site, reinforced – novel, unreinforced) were not considerably damaged, deformation of samples and underlay material was not so big except of underlay of steel plate where perforation of samples occurred in all causes.

Every sample of TPO foils was perforated on each thermal insulation material and little part of sample was ripped out and bored into insulation. In the case of underlay steel plate, little hole made up on a sample.

Unreinforced EPDM foils were not damaged on insulation, but their good elasticity caused damage of underlay insulation by sticking weight puncturing tool into it. The samples were damaged merely on the underlay steel plate.

Reinforced EPDM foils preformed like reinforced PVC-P foils. All samples had been elastically stretched and after that came back to base shape. No damage was occurring except of underlay steel plate where samples were perforated.

Asphalt sheets (coarse-grained) were entirely perforated (2 m drop high) and cracks occurred on the bottom side of samples (1.25 m and 1 m drop high) by puncturing tool on the EPS underlay. Samples placed on the MW and PIR were not damaged but on the steel plate they were.

5 Conclusions

Some certain type of mechanical damage (huge or little) is absolutely not accepted from waterproofing point of view. However, some unfortunate cases could happen (dropping sharp element, nail perforation by mounting some constructions, etc.) in the practise during realization or maintenance of roof. Purpose of this research is to find out dependency between mechanical damage and waterproof function of membranes and after that try to devise range of restoration works. Results showed that certain types of damaged membranes have better ability to withstand water pressure in comparison to each other. It is necessarily for engineers and architects to have knowledge about performance of roof membranes under certain working operations and climatic conditions as well. Workers on the site have to be instructed about the fragility of roof membranes before realization of a construction. Climatic conditions as hailstorm impact could create different performance of membranes placed on the different underlay material. Right selection of underlay is the most important especially for the regions with high dangerous rate of hail storm impact incidence. For the further research is necessary to investigate more membrane samples with combination to other type of mechanical damages and underlay materials.

This research project has been funded by the Ministry of Education, Science, Research and Sport of the Slovak Republic under the contracts VEGA 1/0710/13.

References

1. M. Novotný, I. Misar, S. Šutliak, *Hydroizolace plochých střech. Poruchy střešních plášťů* (Nakladateľstvo STU, Bratislava, 2014)
2. *STN EN 192, Flexible sheets for waterproofing. Bitumen, plastic and rubber sheets for roof waterproofing. Determination of watertightness* (2001)
3. *STN EN 12691, Flexible sheets for waterproofing. Bitumen, plastic and rubber sheets for roof waterproofing. Determination of resistance to impact* (2006)

The problem of the compatibility of bitumen sheets for the reconstruction and rehabilitation of roofs

Jan Plachý[1,*]

[1] Institute of Technology and Business in České Budějovice, Department of Civil Engineering, 370 01, České Budějovice, Czech Republic

Abstract. During the renovation of flat roofs, the waterproofing layer is often repaired locally rather than completely replaced. When making such repairs, it is very important to know whether the proposed bitumen sheets are compatible in order to prevent delamination and the release of the joints. This paper deals with the determination of the resistance of bitumen sheet joints against peeling when combined with different bitumen materials. On the basis of the described test methodology it was possible to determine whether the selected bitumen materials were compatible with bitumen sheets. As part of the experiments, modified bitumen sheets with plastomers (PO), oxidized bitumen (OX), and elastomers (SBS), were tested.

1 Introduction

Many problems associated with roof cladding are related to the release of the bitumen sheet joints. This paper deals with the determination of the resistance of bitumen sheet (AP) joints against peeling [1].

During the renovation of flat roofs, the waterproofing layer is often repaired locally rather than completely replaced. When making such repairs, it is very important to know whether the proposed bitumen sheets are compatible in order to prevent delamination and the release of the joints. In this study, for experimental purposes, bitumen sheets with various kinds of coatings were therefore combined. At present, the coatings on bitumen sheets (AP) are often modified with polymers of either a plastomer or elastomeric nature. This poses a problem with regards to AP joints. From the viewpoint of the chemical composition, it is not always compatible to join sheets with a modified coating made from certain types of polymers of the plastomer type and an asphalt mass modified with elastomeric polymers.

The advice from manufacturers and suppliers with regards to the combination of bitumen sheets (AP) with a modified coating of a plastomer character with other types of coating materials, varies. Some say it is possible, whereas others advise against it or state

* Corresponding author: plachy@mail.vstecb.cz

that it is prohibited. The compatibility depends directly on the type of polymer used in coating.

At present, the identification and labelling of modified bitumen sheets of a plastomer nature is inaccurate. Within this context, the terms APP and SBS require closer attention. SBS is a modification of an elastomeric nature, which is represented by the copolymer of styrene - butadiene-styrene. In contrast, APP is a modification of a plastic nature.

Up to the present day, the use of polymers for SBS modification has remained almost unchanged. The only differences between the individual types of polymers is the proportion of styrene and the diblock number of styrene-budadien. This differs to the situation with regards to the polymers which are labelled as APP. The APP labelling applies to atactic polypropylene, which results in the production of isotactic polypropylene (IPP). At present, APP is only used marginally in production. Polymers from the group of polyolefins (PO) are used as a modifier in the bitumen mass. These polymers are primarily manufactured for the production of modified bitumen. Such modified bitumen is either used for roofs or roads. The mechanical parameters and the quality of the bitumen mass made of PO is higher [2]. The peel strength in polymers of primary chemical synthesis - polyolefins (PO) is comparable to the bitumen mass modified with SBS. In secondary chemical synthesis the peel strength of APP is lower by approximately 30%. Not all manufactures use PO. It is for this reason that some bitumen sheets (AP) of a plastomer nature are problematic and others are not.

A number of authors [2-7] have dealt with the shear and peel resistance of joints. The latest author to do so, produced very similar results [8]. The strength of joints has been compared to AP joints with a coating modified with SBS and sheets with a coating from oxidized bitumen. The strength of the joints of AP with APP were much lower. The strength of sheets with a coating modified with APP, peeled at room temperature, was 25 to 50% of the strength of bitumen sheets with a coating modified with SBS. A principle of ageing bitumen mass with modification SBS is given for an overview [4].

SBS copolymers belong to the group of block copolymers, whereas APP to the group of polyolefins. In the case of SBS copolymers, it forms a mixture with the bitumen, which due to its aromatic form with bitumen (styrene is aromatic in nature) creates a three-dimensional physical network between the polystyrene blocks in the bitumen. Its mechanical properties and stability over time are determined by the three dimensional physical network, wherein the butadiene bond ensures the elasticity of the entire polymer-bitumen mixture. Ageing over time due to exposure to climatic conditions, leads to the decoupling of the bonds and thus to a loss of its elastic properties.

In the case of APP modifications, and polymers in general from the group of polyolefins, the chemical relationship with bitumen is considerably smaller. APP (polyolefins) makes an emulsion with bitumen. Over time, a separation occurs in each phase (between the bitumen and the polymer) which is caused by their different surface tensions. It is also caused by the different molecular sizes of the two phases and their mobility. This separation results in the strength reduction that occurs at the joints.

2 Materials, testing methodology and assessment

For the actual tests six sample combinations were prepared from AP with a modified coating from SBS, PO, or oxidized bitumen (OX). The combinations were as follows: SBS+SBS; SBS+OX; SBS+PO; PO+PO; OX+OX.

The tests were conducted at a room temperature of +23°C and in part at +50°C for capacity reasons and on the basis of previous results [2, 3]. The tests at +50°C were not carried out in a climatic room. The number of samples of the aforementioned combinations was determined to be five in accordance with the methodology in [1]. A frame shift of a load machine is according to [1] to 100 mm per minute. The test samples (TS) were clamped into the jaws of the testing body (see Figure 1). The low numbers of test samples (TS) place very high demands on the preparation of its own TS. The actual tests were carried out after thermal ageing at 70°C after 0 (new), 2, 4, and 8 weeks. The evaluation was only conducted for the maximum force F_{max}.

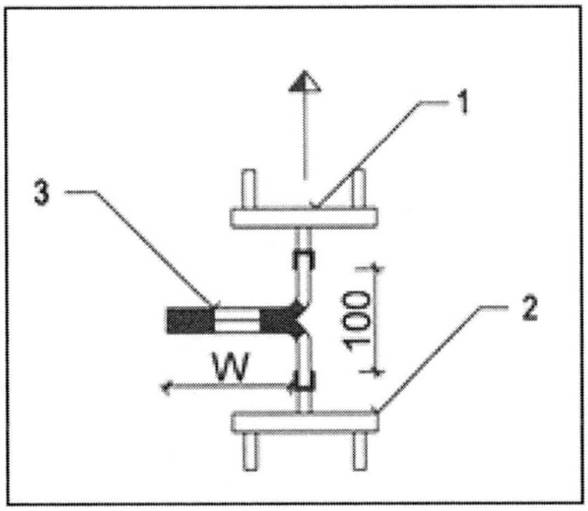

Fig. 1. Strength testing of joints against peeling according to [1]. 1 - frame of testing machine; 2 - clamp of the testing body; 3 - bitumen sheet joint. Source: Author.

5 test samples of the determined combinations were used for testing. The final force is presented in the form $\overline{x} \pm \overline{\vartheta}(\overline{x})$, where $\overline{x}$ is the arithmetic average and $\overline{\vartheta}(\overline{x})$ is the probable measurement error. The latter was determined on the basis of the error theory for deviations in the arithmetic average using the formula:

$$\overline{\vartheta}(\overline{x}) = \frac{2}{3}\sqrt{\frac{1}{n(n-1)}\sum_{i=1}^{n}(x_i - \overline{x})^2}. \qquad (1)$$

3 Results

Table 1. Laboratory measurements of AP joint peel resistance for different types of coating - F_{max} (N/50mm)

Test samples		Mean value of peel resistance of the joints F max (N / 50 mm) at 23°C			
Type of mass bitumen coating	Ageing at 70°C (weeks)	The mean value	Probable error	Standard deviation	Coefficient of variation (%)
SBS+SBS	0	2228	12.0	40.3	18.1
	0 (50°C)	54.9	4.8	16.0	29.2
	2	189.6	5.3	16.0	8.5
	4	218.5	5.9	17.6	8.1
	8	245.2	15.8	41.1	16.8
	8 (50°C)	55.7	1.9	5.0	9.0

SBS+PO	0	121.2	5.7	19.0	15.7
	0 (50°C)	21.9	0.8	2.8	12.8
	2	99.0	4.3	12.9	13.0
	4	98.9	8.0	23.9	24.2
	8	109.2	3.6	10.7	9.8
	8 (50°C)	28.2	2.0	5.3	18.8
SBS+OX	0	40.2	3.3	11.0	27.4
	0 (50°C)	17.5	1.2	4.1	23.4
	2	62.0	5.3	16.0	25.8
	4	60.6	2.6	7.8	12.9
	8	46.1	3.4	10.1	22.0
	8 (50°C)	23.0	0.9	2.4	10.3
PO+PO	0	114.0	11.4	34.1	29.9
	0 (50°C)	24.2	0.5	1.6	6.8
	2	109.6	3.5	10.4	9.5
	4	122.0	6.9	20.6	16.9
	8	141.6	8.2	24.6	17.4
	8 (50°C)	31.4	0.2	0.6	2.0
PO+OX	0	21.4	1.4	4.8	22.3
	0 (50°C)	7.5	0.2	0.6	7.5
	2	25.6	1.9	5.7	22.2
	4	32.7	4.0	12.1	37.1
	8	30.8	1.2	4.1	13.2
	8 (50°C)	9.5	0.3	0.6	5.8
OX+OX	0	32.4	2.1	7.0	21.5
	0 (50°C)	18.1	0.4	1.5	8.2
	2	35.6	2.6	6.8	19.2
	4	34.4	2.5	7.5	21.9
	8	32.5	1.0	3.4	10.4
	8 (50°C)	17.0	-	-	-

4 Discussion

With respect to the ageing process, all the joints for all the combinations gained strength at a temperature of +23°C; the only significant exception is the SBS+PO combination. However, even the strength in this combination increased in comparison to that of OX.

None of the joints lost strength at 50°C. However, the strength of the PO+OX combination is fairly low. The strength gains amounts to 50% of the total strength of the OX+OX joint. This may be considered the critical strength.

The joints with combined strengths at +23°C and +50°C shows results that are comparable with those of the non-combined strengths. The only exception is the SBS+PO

combination at 23°C. No substantial changes occur with regards to the different temperatures and the ageing process of the sample.
Owing to the poor workability of the bitumen sheets with the coating made from oxidized bitumen at 50°C, the PO+OX combination and the OX+OX combination, only two bitumen sheets and one sample were tested.

The final strength of the SBS+PO joint is slightly lower than the strength of PO+PO joint despite the strength of SBS+SBS being ca. 100% higher than the strength of PO+PO. The result showed a slight degradation of the joint.

The final strength of the SBS+PO joint is higher than the strength of OX+OX joint. The result did not indicate any degradation of the joint.

The final strength of the PO+OX joint is lower than the strength of OX+OX joint despite the strength of OX+OX being ca. 25-30% of the PO+PO strength. The result revealed a serious degradation of the joint.

The results of the laboratory testing suggest that the strength of the combination of the OX sheets with the SBS sheets and PO, is low. The peel resistance of the SBS+SBS and PO+PO modifications does not meet the value requirements [2]; however, their values are higher than the results suggested in [3] and [5]. With regards to the plastomer sheets, the procedure [5] included a non-identifiable mass labelled APP. In contrast to the statements of the authors mentioned above, only the SBS+PO combination showed a sharp decrease in strength.

5 Conclusion

The results indicate that a combination of different masses of bitumen is a viable solution for the reconstruction and rehabilitation of roofs. However, consideration should be given to the right combination of the coating of bitumen. The results of the test show that it is possible to combine the SBS modified coating with a coating of oxidized bitumen or a PO modified coating. The combination with the coating from oxidized bitumen is considered hazardous for this kind of polymer because the strength of the joint is considerably lower than the strength of the AP joint with the coating from oxidized bitumen.

References

1. *ČSN EN 12316-1, Flexible sheets for waterproofing - Part 1: Bitumen sheets for roof waterproofing - Determination of peel resistance of joints* (2000)
2. M. Novotný, Střechy, fasády, izolace, 7, 44-45 (1999)
3. J. Plachý, *Spoje asfaltovaných izolačních pásů za působení času a teploty ve skladbě jednoplášťových plochých střech* (Brno University of Technology, Brno, 2005)
4. J. Plachý, *Collection of papers: Building Defects* (2011)
5. R. Šmehyl, *Pevnosť vzájemného spoja asfaltovaných povlakových krytín* (Bratislava, 2007)
6. T. Petříček, *Adheze spojů asfaltových pásů mechanicky kotvených* (Brno University of Technology, Brno, 2013)
7. J. Stodůlka, *Mechanicky kotvené povlakové hydroizolace* (Brno University of Technology, Brno, 2014)
8. M. Novotný, I. Misar, S. Šutliak, *Hydroizolace plochých střech. Poruchy střešních plášťů* (Grada, Prague, 2014)

Differential substrate subsidence of the EnviHUT project pitched extensive green roof

Klára Nečadová[1,*], *Petr* Selník[1] , and *Hana* Karafiátová[1]

[1]Brno University of Technology, Faculty of Civil Engineering, Veveri 331/95, Brno, Czech Republic

Abstract. In primary phase of testing building physical characteristics of the EnviHUT project extensive and semi-intensive roofs with 30° inclination occurred exceptional substrate subsidence. An extensive testing field with retaining geocell-system evinced differential subsidence of individual sectors after six months. Measured subsidence of installed substrate reached 40 % subsidence compared to originally designed height (intended layer thickness). Subsequent deformation of geocell-system additionally caused partial slide of substrate to drip edge area. These slides also influenced initial development of stonecrop plants on its surface. Except functional shortages the aesthetical function of the whole construction is influenced by the mentioned problem. The stated paper solves mentioned issues in view of installation method optimization, selection and modification of used roof substrate and in view of modification of geometric and building installed elements retaining system arrangement. Careful adjustment of roof system geometry and enrichment of original substrate fraction allow full functionality from pitched extensive green roof setting up. The modification scheme and its substantiation is a part of this technical study output.

1 Introduction

In AdMaS sites of Brno University of Technology AdMaS there is ongoing longitudinal study of behaviour of different variants of vegetative roofs. The main test unit is the project EnviHUT, where were after 10 month of existence registered significant volume changes of substrates of extensive green roofs and geometric rearrangement of the restraint system. This paper is about evaluation and description of these two major defects in construction.

1.1 Construction description

1.1.1 EnviHUT

EnviHUT is designed as a small portable holiday building with container frame construction and roofed with pitched green roof with three variants of vegetation layer. Used building materials of individual enveloppe construction were chosen with respect to

* Corresponding author: necadovak@fce.vutbr.cz

the requirements of building-physical measurement of new recycled or local natural materials and construction designs. The EnviHUT is self-sufficient independent object from the perspective of use of energy and its technological operation is provided with associated wind power station, solar roof panels and automated control battery unit [1].

EnviHUT is intentionally exposed to aggressive impact of climate, especially strong wind, therefore many North European design solutions and components were used in construction. At the same time this negative effect was turned into a benefit in the form of high-quality ventilation of ventilated spaces using the chimney effect [2].

The choice of natural and recycled building materials meets the needs of research of each team member. On the roof synthetic partly recycled roofing substrate and local natural soil are installed. Also a vegetation retention layer of recycled polyester bi-component developed by authors is installed.

Structural design in terms of temperature oscillation in the hot summer was taken into consideration. Based on the quantity of the soil [3,4], rich vegetation, vegetation-associated retention panels [5] and the design of the roof with sufficient overlap the final design oscillation frequency of the temperature was set as 12 hours. With this proposal, in combination with a controlled ventilation on the north side, should be ensured optimum internal temperature environments and surfaces, without requiring the operation of energy-intensive air conditioning units.

1.2 Assessed test segments

The first segment has structure formed from sheets of recycled PES with a thickness of 60 mm, which form a vegetation-retention layer, also cellular capture system, which is filled with crushed substrate with the planned thickness of 120 mm, and finally formed by vegetated sedum plants. This system represents a roof with poor vegetation.

The middle segment is based on the Swiss concept and represents a biodiversal roof alias a roof with rich vegetation. This composition is supported with vegetation retention-layer boards made of recycled polyester 40 or 60 mm thick, on which is located a restraining system of a plate material and a reinforcement mesh. On this layer is stacked 180 mm of soil from excavation. For speeding up growth vegetation grass mats complemented by the seeds of surrounding plants (poppies, various grasses, camomile etc.) were used.

The last segment represents an Icelandic concept of pitched green roofs realization. On the vegetation-retention layer of different thickness are laid grass mats between which is installed reinforcing net and natural soil. The first two mats are laid with roots up and the last mat, which forms the surface structure, is laid with roots down. This way designed and implemented layer forms very resistant and stable system that can resist extreme weather.

Consolidation in the Nordic and semi-intensive biodiversity variant was successful. Slides in monitored area were not registered. Subsidence went equally in whole area and orientation.

1.3 Vegetation retention layer

Installation vegetation-retention plate with a total dimension of 1.2 x 0.8 m was sticking this layer to a securing waterproofing layer in the longitudinal direction. (Longitudinal direction was chosen in order to reduce mutual joints of entire layer.) Surface of securing waterproofing layer had to be cleared off layers of polyethylene – burned down. The adhesive low-expansion polyurethane foam which was applied after 20 to 25 cm in the vertical direction across the width of the plate was used. The reason for using low-expansion polyurethane foam was its higher mold resistance, limiting fungus growth in the connection and high adhesion to the connected surface. Glued mattresses were temporarily for 2 min weighted down so that the equalization occurred in their entire length and formation of joints there between them was reduced. The joint between the mattresses was not glued together in order to maintain uniform flow characteristics of the entire cross section of the layers in the structure. Details, corner and edge parts were then formed, so as to ensure homogeneity and integrity of layer [3].

Due to the different thickness of the used mats a model of the amount of water collected in monitored segments can be made [6]. On EnviHUT were used two types of mattresses, namely:

- Mattresses about 4 cm thick in the performance of 4000 g/m^2,
- Mattresses about 6 cm thick in the performance of 6000 g/m^2.

A thicker version was used on the whole area of the south side of the roof and at the ridge of the northern exposure with a width of 80 cm. On the rest area was glued a mattress with a smaller retention capability. After gluing this layer was left for 12 hours without any further intervention. On this layer the installation of the catching system, the substrate and the layers of vegetation followed.

1.4 Catching system and planting of vegetation

The installation of geocells net was anchored to an additional frame element at the ridge and on the sides of the roof. Partially (20 % of the length of the ridge) is the net also folded over the ridge. Subsequently, the area was backfilled with original roof substrate (by a Czech producer) in direction from the ridge to the eaves.

This installation was not realized directly by the authors of the project. It was entrusted to the company, which realize vegetative roofs on a long-term basis.

The chosen form of installation did not allow more efficient compaction. After installation that part showed large volume changes caused by uneven consolidation of the dependent variable shape of each connected segment. Since the installation was made before the dormancy period of planted vegetation, it has not grown fast. The contact surface was exposed to mild winter conditions in the following months. At the end of the winter small surface landslides were observed. Caused landslides were observed on both of installed exposures (NNE) – (SSW). At least 5 of the 25 applied packages (40 kg) contained uncrushed expanded clays and shales.

2 Measurement conditions

The main test surface is gabled green roof with an inclination of 30° with NNE - SSW orientation and testing segment with the size of 2.4 x 2.5 m. Gabled roof structure has an inclination of 30° and is divided into three sections.

Installation of the roof took place at the end of September and during October 2015 with adapted vegetation-retention plate of polyester fibers and clusters with the performance of 4000 g/m^2 at a height of 4 cm and 6000 g/m^2 at a height of 6 cm. Subsequent winter was very mild with little precipitation, without extreme freezing season. The temperature in the monitored period fell only one day under the designed - 12 °C. The only one month with a negative average temperature was January -1.94 °C. Not divided installed geocells were measured in the vertical direction v and horizontal direction h. Subsequently, the mutual evaluation of the measurement dimensions individually to the whole were done. Divided cells were not included in this evaluation. Accuracy of measurement was set to ± 5 mm, what is twice the width of the installed geocells net. Installation average dimension v_i x h_i was measured as (40 ± 0.5) x (40 ± 0.5) cm.

The aim of the evaluation is to define the degree of the creep after the consolidation of the structure and also to evaluate the question of uniformity subsidence of studied element.

The measurement results from the May 30th, 2016 were systematically noted in Table 1 for the south side and in Table 2 for the north side. The table is not the standard form of record but more chart from which the final forming of entire restraint geo-cell system demonstrated in Figure 1 can be seen. Some cells were not depressed on the surface. These cells could not be measured without disturbing the existing roof surface. In the table, this condition is marked with the term *cov* as covered. Some cells were so deformed that it was impossible to measure monitored dimension. These cells are in table marked with the term *dam* as damaged.

Fig. 1. A surface at eaves on the northern side.

3 Results

Table 1. Final shaping of geocells on the southern exposure.

Final shaping of geocells on the southern exposure												
[cm]			Vertical line									
			I		II		III		IV		V	
Horizontal line	J			v								
		I	41	h								
	H		46	51		52		54				
		G	45	53	cov	48	54	44,5	53		cov	
	F		54	cov	cov	51	44,5	49	38,5	44	cov	
		E	41,5	cov	48,5	45,5	51,5	50	50	53,5	40	
	D		54	47	50,5	48,5	48,5	48	48,5	46,5	46,5	49
		C	42	52	47	50	43,5	50	47,5	52,5	dam	47,5
	B		51	42	51,5	42,5	55	44,5	50	51,5	dam	
		A	46,5	55,5	38,5	54	41	53,5	46	43,5	52,5	
			48,5		58,5		55,5		46		35,6	

Table 2. Final shaping of geocells on the northern exposure.

Final shaping of geocells on the northern exposure												
[mm]			Vertical line									
					II		III		IV		V	
Horizontal line	J											
		I	v		52		52		47		43	
	H		h	43,5	46,5	52,5	46,5	46,5	52	46	55	
		G		52	45,5	46	46,5	52	46,5	52,5	49	
	F			cov	51	40	53	47	52	50	49	
		E		cov	43,5	57	42	51	49	50	41	
				50	54,5	43,5	57	45	48,5	41	40	
	D	C		44	50	53,5	45,5	53	37,5	56	42	
				35,5	46	51,5	53,5	38	59	40	52	
	B	A		47	36	46	50	59	38	55	52	
					60		52,5		53		47	

3.1 Measurement comment:

3.1.2 The South side

- The diameter of the unit v = 46.95 cm; h = 48.10 cm,
- Series G - misalignment in subsidence 1 to 4 in the horizontal direction,
- Series E - misalignment in vertical direction E1 to E2 and also E4 and E5,
- Series D - balanced settlement,
- Series B - B4 significant vertical deformation against the line and also the uneven settlement to horizontal direction,
- Series A - a drop in columns A2 followed by a significant horizontal deformation A2 and A3,
- Sum of columns 2,3,4 (194; 190; 196.5 cm) is about 18.5 cm different compared to column 1 (175 cm),
- The sum of the rows E, D, A (248, 252, 244 cm) - uniform subsidence in the horizontal direction relative to the whole.

3.1.3 The North side

- The diameter of the unit v = 45,09 cm; h = 51,49 cm,
- I2, I3 are vertically mutually equivalent; I5 is shifted against previous units,
- The last column G Series - partly compensates changes caused by the series I,
- Series D and column 1 is significantly deformed to the whole line,
- The most significantly deformed is unit C4 and its surroundings,
- A2 and C4 is the most deformed at all,
- B3 and B2, F2 and F3 – is misalignment offsetting rows B and F,
- Sum of columns 3, 4 (236, 218 cm) is different from the outer column 2, 5 (227 cm),
- The horizontal sum of rows A, B, C, D, E, G, H, I (212.5; 207; 210.5; 206.5; 200; 205; 202.5; 200) - Part E up to I is uniform in horizontal plane, followed by an increase in the ranks of D and following similar lines a to C.

3.2 Proposed changes of the installation procedure

The substrate is a key aspect that the most affects the stability of the final installation. Especially for substrates with a component of expanded clays and shales. Within supply of these materials must be strictly checked actually delivered substrate and verify the main required installation parameters - these parameters are mainly grain composition and crushing level of solid particles. Here is the critical point of substrate purchase and application. Some substrates are limited by technological possibilities of mixing and subsequent packaging according to the forms of distribution. Unsuitable packaged substrates with less crushed component, which is getting to the substrates by mixing light particles and by insufficient output control, in the realization must be expected and consistently conduct its own process control of applied material. The unfavorable combination of those processes can result in landslides as well as test implementation demonstrated in Figure 3.

This problem could be compensated by own adjustment of the substrate composition, especially to enrich provided substrate with the other bigger fractions. This process was consulted with experts from the BOKU who recommend to enrich the existing installation with stabilizing binding fraction (16-32 mm). Mentioned fraction would not be no longer a part of roofing substrates according to FLL Guidelines, but this is correction of an existing layer. The reason is to create a granularity step curve, so that it can lead to the formation of bonds which prevent landslides on pitched roofs. This corrective application is the most effective measure from the perspective of the authors and collaborating institutions.

Fig. 2. A landslide at the eaves on the south side.

4 Conclusion

Expansion of pitched green roofs applications is connected to the quality of implementation examples that can reach potential customers. Aesthetic and functional defects, which are monitored on a test object, must be preceded by strict implementation of the technological process. The mentioned study points out the critical points of used forms of implementation.

Due to large deformations of individual geocell and the whole restraint system the team of authors recommends to change the size of the grid applied geocells from 40 x 40 (H x D) to half in both directions. Individual partitions of geocells are overburdening with the substrate. In conjunction joints it consequently comes to dependent changing of the entire grid of restraint system. Defect can be caused also by increased local weather effects on a small contact area. This leads, in combination with the already discussed problem of substrate supply, to landslides. These landslides secondarily damage the geometry of system and do not allow continuous development of surface vegetation. The exposed areas are degraded by wind erosion and partly also rain helps to float the substrate. This erosin on the tested object is enhanced by the large amount of installed spherical expanded beads, which successively reach the contact surface. Here are due to its low weight affected with wind and accumulate at the eaves of the roof.

Corection of such roof is very expensive particularly in terms of time and it would be possible to avoid it with strict control of used substrate before installation. Declared material supply in our case did not match the required characteristics in 20 % of its volume. Especially crushing expanded beads was critical.

The damaged layer will be repaired in October 2016 and observe with the same geometry of the restraint system. Further measurements with the same parameters will be done before this modification.

Authors would like to express their appreciation and thanks to Dr. Bernhard Scharf and Dr. Ulrike Pitha, their research team and the support from the Universität für Bodenkultur Wien and to Dr. David Bečkovský and Dr. Martin Mohapl from the Brno University of Technology.

This paper was completed under project no. LO1408 "AdMaS UP – Advanced Materials, Structures and Technologies", supported by the Ministry of Education, Youth and Sports under the National Sustain- ability Programme I and under the Project FAST-S-16-3345 supported by Ministry of Education, Youth and Sports under "Specific University Research".

References

1. T. Pereira, et al., *The 1st European Urban Green Infrastructure Conference* (Institute for Sustainable Engineering, University of Palermo, Vienna, Austria, 2015)
2. G. Kotsiris, A. Androutsopoulos, E. Polychroni and P. A. Nektarios, Energy and Buildings **45**, 240-249 (2012)
3. T. Emilsson, *Ecological Engineering* **33**(3-4), 265-277 (2008)
4. K. L. Getter, D. B. Rowe, *HortScience* **44**(2), 401-407 (2009)
5. J. Mentes, D. Raes, M. Hermy, *Landscape and Urban Planning* **77**, 217-226 (2006)
6. N. D. Vanwoert, D. B. Rowe, J. A. Andresen, C. L. Rugh and L. Xiao, *HortScience* **40**(3), 659-664 (2005)

The analysis of implementation options of transverse joints of bitumen sheets

Jan Plachý[1,*]

[1] Institute of Technology and Business in České Budějovice, Department of Civil Engineering, 370 01, České Budějovice, Czech Republic

Abstract. One of the most frequent sources of faults on flat roofs is a violation of the waterproof layers along the transverse joints between bitumen sheets. This fault is caused by a failure to adhere to the technological methodologies. This paper presents current solutions, as well as a new solution, for creating such joints. The new solutions are experimentally compared with existing solutions.

1 Introduction

Adhesive bitumen sheets (ABS) represents an up-to-date and sophisticated technology. ABS technology has huge advantages, as well as severe deficiencies. The advantages are [1]:

- its practical application,
- after application it is strictly prohibited to work with an open flame; the technology is therefore suitable for:
  - waterproofing premises that are not at great risk of catching fire,
  - connecting bitumen sheets to constructions without the need for welding (skylights, barrier glands),
  - laying on polystyrene foam in comparison to assembled panels,
- the flatness with which the bitumen sheets can be laid thereby preventing unequal strain (welding overlaps),
- its economical application in terms of gas (welding torch) or electricity (hot-air welding).

The disadvantages are [1]:

- its application depends on climatic conditions,
- it is not possible to inspect the joints in detail (the bitumen cast is missing),
- it is not possible to guarantee that the transverse overlaps are fully waterproof; most ABS systems include gritting; the sheets must therefore be flame-welded or hot-air connected.

The subject of this article, as described in brief below, is part of ongoing in-depth research. The main goal of this article is to generate a transverse joint without the need for thermal treatment. Whereas the application of the majority of bitumen sheets requires thermal treatment, either flame welding or hot-air drying, to produce transverse joints, this does not apply to bitumen sheets without gritting. However, if the length of the applied

* Corresponding author: plachy@mail.vstecb.cz

bitumen sheet is different from the adhesive bitumen sheet from the manufacturer, it remains necessary to apply proper thermal treatment. A number of viable options are put forward to tackle this crucial issue. These options are subsequently subjected to strength tests and compared to the traditional solutions.

2 Material and methods of testing

2.1 Materials

ABSs can be divided into several groups according to their exact composition. In practice, there are three common ABS compositions.

Both the upper and lower cover material consists of bitumen material of the same adhesiveness. These types of adhesive bitumen sheets are widely manufactured. In terms of production costs, lower quality polymers are usually preferred; high quality polymers are used sparingly. For this reason, the bitumen sheets are characterized by a lower adhesiveness at the base in contrast to other adhesive sheets. However, this type of bitumen sheets may be better when it comes to creating transverse joints because the upper and lower cover materials fuse better.

The upper cover material consists of common polymer-bitumen material; the lower cover material consists of a highly adhesive bitumen material. This type of adhesive bitumen sheet is produced by those manufacturers which have the capacity to apply two different kinds of material to the bearing batting at the same time. The application of the common polymer-bitumen material as the upper cover permits the application of highly adhesive bitumen material made from special polymers as the lower cover without the price of the product rising. In addition, the high adhesiveness of these products is preserved.

The upper cover material consists of common polymer-bitumen material; the lower cover material consists of common polymer-bitumen material covered by a thin and highly adhesive bitumen material. In order to produce this type of adhesive sheet, it is necessary to employ white-hot technology. With respect to cost, this option is the most effective. However, there are minor issues with regards to its production and the long-term functionality of the highly adhesive thin layer prior to application.

For the experimental part of this paper the following materials were selected:

1) ABSs with the upper and lower cover material consisting of the same highly adhesive bitumen material; the bearing batting PES + ST 250 g/m^2 with fine grain mineral gritting; flexibility at low temperatures according to ČSN EN 1109:2009 is -15 °C;

2) Double-sided ABSs with heavily modified bitumen material;

3) Single-sided ABSs for sealing; with heavily modified bitumen material with external coating in the form of Al foil and geotextile.

2.2 Methods of testing

The peeling resistance of the formed transverse joints was tested in accordance with ČSN EN 12316-1 [2] (see Figure 1), as was the shearing resistance in accordance with ČSN EN 12317-1 [3] (see Figure 2). With respect to those ABSs with an upper external coating made of fine grain gritting, the size of the transverse joints was fixed at 100 mm.

DOI: 10.1051/matecconf/20179302006

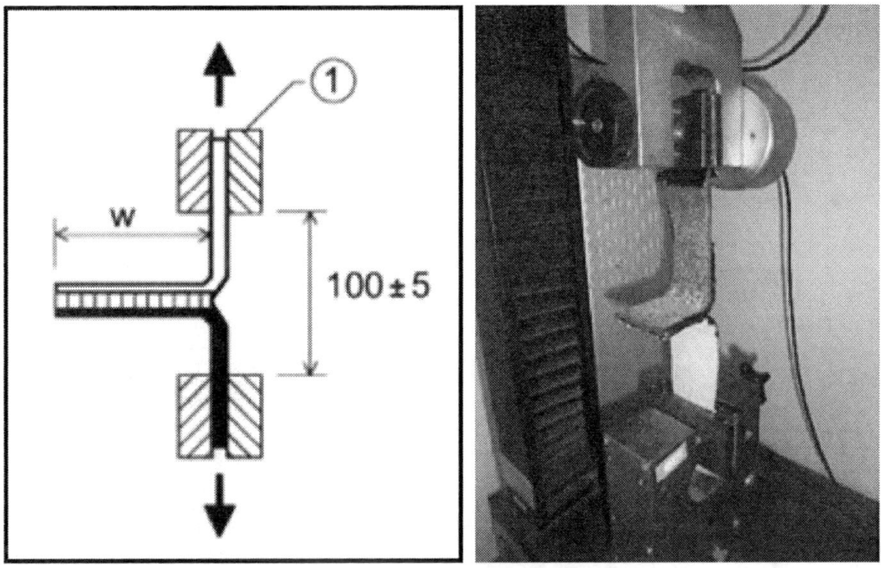

Fig. 1. Diagram of joint peeling resistance test in accordance with ČSN EN 12316-1. Source: author.

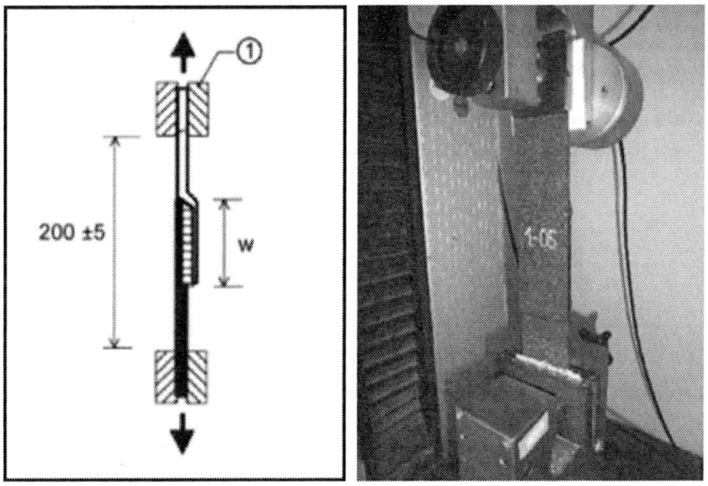

Fig. 2. Diagram of joint shearing resistance test in accordance with ČSN EN 12317-1. Source: author.

Two viable options for creating transverse joints were analysed:
1) Taping with double-sided ABSs of 75-100 mm width without ABS base adjustment. The ABS is firmly taped to the fine grain mineral gritting (see Figure 3);
2) Pushing the parts together and sealing it with an ABS of 100 mm width (see Figure 4).

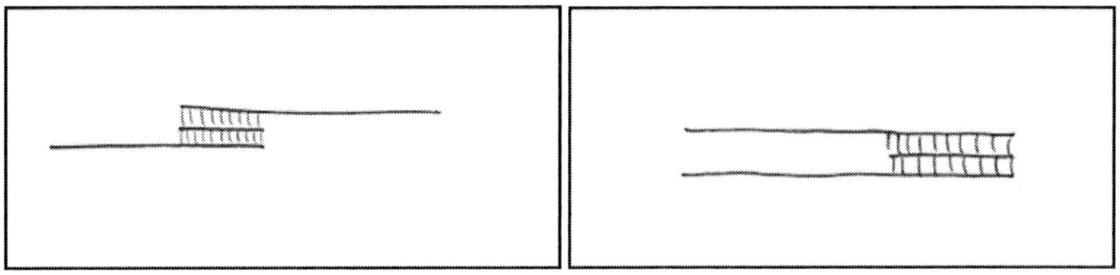

Fig. 3. ABS transverse joint taped with a double-sided adhesive bitumen sheet. Left – shearing resistance test sample; Right – peeling resistance test sample. Source: author.

 DOI: 10.1051/matecconf/20179302006

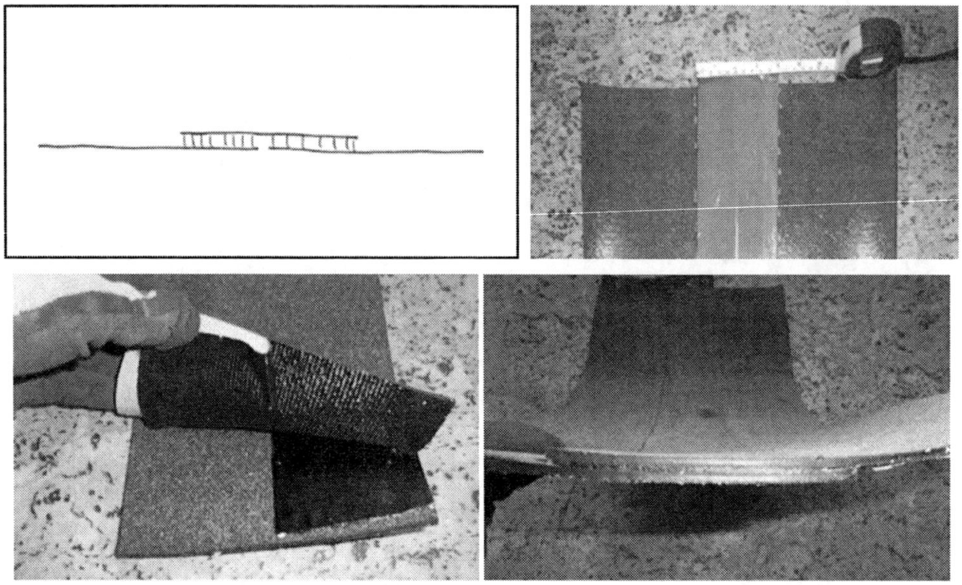

Fig. 4. ABS transverse joint created by pushing the parts together and sealing it with an ABS. Shearing force testing sample. Source: author.

The results of the strength tests according to [2] and [3] were compared to the strength of those joints created using thermal treatment i.e. flame welding (see Figure 5) and hot-air welding (see Figure 6).

Fig. 5. Creation of transverse joints of 100 mm width by flame welding. Source: author.

 DOI: 10.1051/matecconf/20179302006

Fig. 6. Creation of transverse joints of 100 mm width by hot-air welding. Source: author.

3 Results

The transverse joints were tested for their adherence to the tensile strength of the ABS which was determined as 734 N/50 mm. The joints created by welding achieved 81–87% of the ABS tensile strength. The joints created using double-sided ABS achieved approximately 65% of the ABS tensile strength. The joints sealed with ABS achieved 14% of the ABS tensile strength.

Table 1. Test results for the shearing resistance and peeling resistance of transverse joints.

Transverse seal (front)	**Tensile strength [N/50 mm]**	**Shear resistance of joints [N/50 mm]**	**Peel resistance at the joints [N/50 mm]**
Standard	EN 12311-1	EN 12317-1	EN 12316-1
Reinforced bitumen sheets [N/50 mm]	734	o	o
Joint design width of 100 mm by melting by fire	o	639	31
Bonding width of 100 mm by melting using hot air	o	597	44
Bonding with double-sided adhesive tape in width of 75 mm	o	482	15
Bonding with double-sided adhesive tape in width of 100 mm	o	475	21
Performing connections to the event and strip bound with tape in widths of 100 mm	o	102	o

4 Discussion

4.1 Determination of shear resistance of joints

After a typical procedure (immersion coating), the joint strength did not achieve the required ABS tensile strength; there was a tolerance of approximately 10% tolerance in the average value for both operating procedures (flame welding, hot-air welding).

With respect to sheet strength, the joint strength after taping with double-sided ABS was insufficient; however, if the absolute value increases to 500 N/50 mm this is considered sufficient. It is also necessary to take into consideration that the joints were created at a temperature of 15°C. As a result, in practice it is not only necessary to apply the appropriate downward force, but also to maintain a steady temperature, which must not rise above 15°C.

The joint strength after sealing achieved only 14% of the ABS tensile strength. The strength of this type of joint is determined by the strength of the bearing batting – AL foil or geotextile. The joint not only proved insufficient with regards to the strength tests, but also as a T joint.

The results must be added to the high-grammage bearing batting with great strength, the complete testing samples implementation and, above all, bitumen material temperature (flexibility at temperature -15 °C)

In order to determine whether the joint with the double-sided tape met requirements, it was put through a waterproof test [4]. The results show that it indeed met the requirements for roofing i.e. 10 kPa. The joint sealed with adhesive tape (pushed together method) was not put through a waterproof test.

4.2 Determination of peel resistance of joints

The peeling resistance test generated low strength values for all the joints. The results were therefore not comparable with the results of the ABS strength tests. The results are only comparable with the different methods for creating transverse joints. The joint strength of the joint created with double-sided ABSs amounted to 30 – 45%. Although the resistance value is not high, it is considered sufficient. This type of joint must have a value of 150N/50 mm. The actual value was 20% higher.

A similar series of tests needs to be carried out with a larger number of samples than is suggested in [2] in order to eliminate the human factor (in terms of the preparation of the samples).

5 Conclusion

The experimental tests show that it is possible to create transverse joints without the need of thermal treatment. Adhesive bitumen sheets offer an effective solution. However, this solution applies only to the base ABS; whereby the upper welded bitumen sheet seals the waterproofing layers.

References

1. J. Plachý, *Střechy 2015: sborník příspěvků* (2015)
2. *ČSN EN 12316-1, Flexible sheets for waterproofing - Part 1: Bitumen sheets for roof waterproofing - Determination of peel resistance of joints* (2000)

 DOI: 10.1051/matecconf/20179302006

3. *ČSN EN12317-1, Flexible sheets for waterproofing - Part 1: Bitumen sheets for roof waterproofing - Determination of shear resistance of joints* (2000)
4. *ČSN EN 1928, Flexible sheets for waterproofing - Bitumen, plastic and rubber sheets for roof waterproofing - Determination of watertightness* (2001)

MATEC Web of Conferences **93**, 02007 (2017) DOI: 10.1051/matecconf/20179302007
Building Defects 2016

Analysis of failures of waterproofing cladding layers of terracotta tiles

Milan Koláčný[1,*]

[1]CTU in Prague, Faculty of Civil Engineering, Thákurova 7, 166 29 Prague 6, Czech Republic

Abstract. The article is focused on one of the most important roof cladding layers – the waterproofing cladding layer of terracotta tiles. Its detailed analysis covers the main waterproofing cladding layers in terms of their material characteristics and installation methods. The article concludes by formulating principles for the correct design of the main waterproofing layer/construction.

1 Introduction

Roof cladding has been part of the main waterproofing construction in various shapes since the beginning of our era. Around the 1970s – 1980s, the attic space (usually of existing residential houses) started to be used for living or other purposes involving permanent people's presence.

One of the objectives of research was to formulate the principles for the effective design of waterproofing layers in roof cladding systems and prove the effect of the main waterproofing layer on the life span of the whole roof construction [1].

2 Experimental verification of reliability and durability of waterproofing cladding layers

The analysis of the mechanical and physical characteristics of waterproofing cladding layers was made on two reference houses with residential attics. The first house had a wrongly installed roof cladding without an effectively ventilated air layer and the roof tiles were locally damaged, the cladding age is 18 years, the slope of roof planes 25°and 32°.

The second reference house had a faultlessly installed roof covering with fired clay roof tiles in accordance with regulations in force ten years ago with a slope of 40°.

Experimental testing covered waterproofing cladding layers installed pursuant to ČSN standards in force and manufacturer's assembly principles, including the correct workmanship of the ventilated air layer. The waterproofing cladding was subjected to impermeability and tensile bending strength tests. This experiment is focused on the properties of materials embedded in the roof envelope in direct connection to their life expectancy and mechanical and physical characteristics.

* Corresponding author: milan.kolacny.rbe@gmail.com

Roof tiles were sampled from the waterproofing cladding layer of both houses and successively subjected to standard tests. To compare the characteristics, roof tiles of the original batch stored in the interiors of both houses were also tested. These roof tiles were tested for impermeability under ČSN EN 539-1 [2], testing method No. 2 and for tensile bending strength under ČSN EN 491:2012 [3].

2.1 Mechanical and physical characteristics of waterproofing layer materials on experimental houses

The results of the tests of mechanical and physical characteristics for differently installed roof tiles from two experimental houses "Trojanova" (without a ventilated air layer) and "U Smaltovny" (faultless roof cladding) are presented in Tables 1 and 2:

Table 1. Evaluation of the impermeability test of plain roof tiles.

IMPERMEABILITY TEST - Trojanova				**IMPERMEABILITY TEST – U Smaltovny**			
Date of testing: 28. 12. 2015 from 5.00 a.m.				Date of testing: 13. 1. 2016 from 8.30 a.m.			
Specimen No.	Weight [g]	Time to first drop [hours]	Impermeability coefficient	Specimen No.	Weight [g]	Time to first drop [hours]	Impermeability coefficient
Roof tiles sampled from a cellar deposit				**Roof tiles sampled from a cellar deposit**			
N1	1674.2	2.25	0.888	N1	1577.1	no drop	0
N2	1678.5	5.5	0.725	N2	1571.8	no drop	0
N3	1675.5	5	0.75	N3	1567.5	no drop	0
N4	1674.9	5	0.75	N4	1570.1	no drop	0
Mean values		**4.44**	**0.78**	**Mean values**			**0**
Roof tiles embedded in the roof envelope				**Roof tiles embedded in the roof envelope**			
S1	1764.9	no drop	0	S1	1631.7	no drop	0
S2	1686.8	no drop	0	S2	1654.6	no drop	0
S3	1764.5	no drop	0	S3	1626.3	no drop	0
S4	1772.5	no drop	0	S4	1624.2	no drop	0
Mean values			**0**	**Mean values**			**0**

In the roof tiles sampled from the house with a cladding without a ventilated air layer, no leakage occurred in any of 4 embedded (used) roof tiles during the testing time, i.e. 24 hours. So-called "tightening", i.e. closing of pores occurs in the used roof cladding and, successively, the roof cladding does not leak during its life span. In the roof tiles from the deposit, which had not been exposed (embedded), the average impermeability coefficient measured from 4 specimens reached a value of 0.778.

Laboratory tests of the roof tiles from the second reference house manifested that the roof tiles which had been embedded in a correctly installed main waterproofing cladding layer as well as the roof tiles stored in the deposit had not lost their principal property – impermeability during the 10-year period.

Table 2. Tensile bending strength of roof tiles.

RH Trojanova, Prague 2				RH U Smaltovny, Prague 7			
TENSILE BENDING STRENGTHS OF ROOF TILES - Experiment 1 and 2							
Specimen No.	Weight [g]	Hanging length [mm]	Max. force F [N]	Specimen No.	Weight [g]	Hanging length [mm]	Max. force F [N]
Roof tiles embedded in the roof envelope				**Roof tiles embedded in the roof envelope**			
1	1673.6	352	792	1	1631.7	336	1173
2	1680.8	351	913	2	1645.6	337	1168
3	1675.5	352	886	3	1626.3	336	1168
X	-	-	-	4	1624.2	336	993
Mean value	**1676.63**	**351.67**	**863.67**	**Mean value**	**16334.2**	**336.3**	**1126**
Roof tiles sampled from a cellar deposit				**Roof tiles sampled from a cellar deposit**			
X	-	-	-	5	1577.1	349	1100
4	1765.4	349	1100	6	1571.8	351	1158
5	1687.7	351	1158	7	1567.5	348	1229
6	1763.2	348	1229	8	1570.1	355	1346
Mean value	**1738.77**	**349.33**	**1162.33**	**Mean value**	**1571.6**	**335.8**	**1304**

Laboratory testing verified a decrease in the tensile bending strength of 3 specimens embedded for 18 years over an unventilated air layer by nearly 25% compared to 3 specimens of identical roof covering stored in the deposit of spare roof tiles on the first overground storey of the same house for the same period of time. On the contrary, 4 specimens embedded over a ventilated roof layer showed a decrease in the tensile bending strength of 14% after 10 years against the specimens from the deposit. In both reference houses, however, the minimum standard strength of plain terracotta roof tiles required for the whole life span of 600N was observed with a reserve.

3 Analysis of failures and defects of roof cladding envelopes

The experiments included the assessment of the causes of failures and defects of roof cladding envelopes. Based on long-term statistics of repairs and reconstructions of 2011 – 2015, the results presented in this study are as follows (internal statistical data of rbe Koláčný s.r.o. Company) are presented in Tables 3 and Fig. 1:

Table 3. Causes of roof reconstructions over a period of 5 years.

STATISTICS OF RECONSTRUCTIONS OF ROOFS with terracotta and concrete roof tiles												
	year / number of roofs											
Cause of repair	2015		2014		2013		2012		2011		Total	
Roofs in total	42	%	56	%	38	%	33	%	46	%	215	%
Additionally built-in attic space	15	35.7%	21	37.5%	8	21.1%	3	9.1%	12	26.1%	59	27%
Wrongly installed under-roof / poor workmanship	13	31.0%	10	17.9%	9	23.7%	2	6.1%	4	8.7%	38	18%
Excessive age of roof covering	14	33.3%	25	44.6%	21	55.3%	28	84.8%	30	65.2%	118	55%

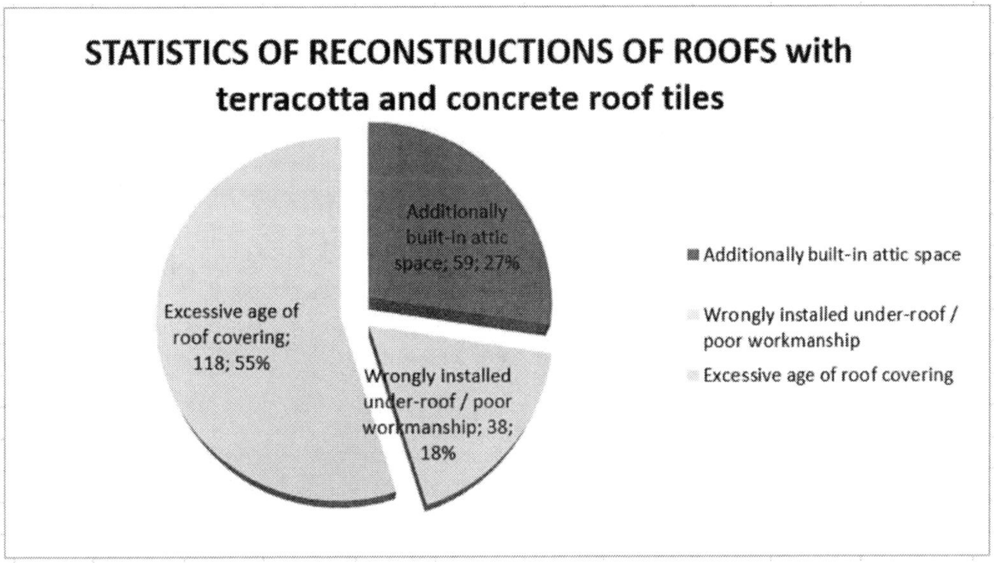

Fig. 1. Causes of roof reconstruction over a period of 5 years – comparison.

Note: The age considered excessive is a period exceeding 60 years.

The above statistics clearly show the growing trend in repairs and reconstructions due to wrongly installed roof cladding or using unsuitable materials. These are, in particular, poor-quality foils for additional waterproofing layers, which are the most frequent cause of a necessary reconstruction.

4 Principles of effective design of main waterproofing constructions in roof cladding systems

These are the principles for the installation of waterproofing constructions of roof envelopes and their effect on the roof construction's life span. The determination of the design life expectancy of the main waterproofing construction requires respecting Instruction F (directive of building products 89/106/EEC) and Eurocode ČSN EN 1990. The above documents recommend a design life expectancy of constructions for common buildings replaceable with some effort of 25 years, it is the life expectancy of easily replaceable constructions. The recommended life expectancy of repairable constructions is

10 years. Therefore, according to this directive, the additional waterproofing layer should have a required life expectancy of 25 let, whereas the life expectancy of 10 years would suffice for the roof tiles. If the manufacturers of additional waterproofing layers declared a life expectancy lower than 10 years, i.e. significantly shorter than the guarantee provided by the manufacturers of roof covering systems, the guarantee for the roof covering could not be fully used.

The above facts clearly show that the life expectancy of additional waterproofing layers in the roof construction in terms of the life span of roof envelopes with a waterproofing cladding layer over residential attics is of principal importance both in terms of the roof surface area and detail.

An ideal solution in the environmental and economic perspective is to unify the durability of the main and additional waterproofing layers within the principal waterproofing construction to the maximum extent possible. This is the optimum in terms of the roof cladding cost-effectiveness.

5 Conclusion

The following conclusions can be drawn from the above facts for an effective design and installation of roof cladding systems:

- unify the design life span of the products of the main waterproofing construction both over the surface area and in detail,
- change the approach to the design of additional waterproofing layers (quality of material solution, change in loading, elimination of negative mechanical, UV and chemical effects),
- search for materials with long-term reliable characteristics for additional waterproofing layers.

A faultless and long-term functional principal waterproofing construction undoubtedly necessarily requires a suitable structural design of the whole roof cladding and the right choice of individual materials for each layer. The materials must be mutually compatible and their planned life expectancy should be identical or at least similar. Correct workmanship during the installation is the last condition so that the principal waterproofing construction is fully functional from the very beginning. Only if the above conditions are fulfilled, a long-lasting and perfect roof can be achieved.

References

1. M. Koláčný, The Effect of Roof Cladding on the Function of Waterproofing Layers (Unpublished)
2. *ČSN EN 539-1, Fired Clay Roof Covering – Determination of Physical Characteristics, Impermeability Test* (2013)
3. *ČSN 491:2012, Concrete Roof Tiles and Trim Tiles for Roof Covering and Wall Cladding – Testing Methods* (2012)

Pathological changes waterproof membranes and their identification

Marek Novotný[1,*] and *Ivan* Misar[2]

[1]Faculty of Architecture, Czech Technical University, Thákurova 9, 169 00 Prague 6, Czech Republic
[2]A.W.A.L.,s.r.o., Eliášova 20, 160 00 Prague 6, Czech Republic

Abstract. With the development of waterproofing systems based on synthetic membranes, especially mPVC, there is an increase in pathological changes of these membranes. Competing manufacturers and pressure on price of the final product have resulted in usage of unstable or low quality raw materials. After production, these materials keep their specific characteristics, however, their aging is accelerated, i.e. they lose their essential qualitative characteristics very quickly. Very important is the detection of these processes which is possible using chemical analysis (gas chromatography, IR analysis and others) as well as using visual inspection of surface by microscope, including 3D modeling. Very important are weldability tests which identify glass transition temperature of the material. The results can then be used as a basis for correct identification of problems with weldability, nonweldability of synthetic membranes, respectively, as well as for identification of normal or abnormal aging of these materials.

1 Introduction

The price war has already broken out in the building industry. Waterproofing materials based on plasticized PVC, shortly mPVC, are no exception. The polyvinylchloride (PVC) is a rigid polymer which must be mixed with the plasticizing agents. The mPVC foil is produced by a mix of polyvinylchloride, plasticizers, stabilizers and other admixtures [1]. The investors are ultimate victims, who must deal with complaints, and reduced durability of waterproofing materials, especially in roof constructions. This situation does not only concern to foil waterproofing materials, but also waterproofing materials base on asphalt.

The foils of mPVC as a significant part of the insulation system in large hall buildings are currently the dominant technical solution. For that reason, the manufacturers will focus on this specific type of waterproofing material. The price is the major key to success. It is necessary to emphasize; the price has a direct relationship on the durability of the material.

The problem of shortened durability does not arise immediately after its production, but after incorporation into the building constructions. This condition is characterized as accelerated aging of waterproofing materials of mPVC type. This phenomenon is caused by

* Corresponding author: novotny@awal.cz

 DOI: 10.1051/matecconf/20179302008

migration of plasticizers from film material. Aging does not have a linear character. The aging is accelerated with respect to increasing surface of evaporation (Figure 12).

2 Materials testing

The consequences are a reduction in frictional resistance against any stress of material, respectively flexibility, at low temperatures, a related tests filler content, respectively material thickness in cross-section. The aging decreases due to the content of plasticizers, thus it is increasing the proportion of fillers, which are not lost, respectively reduction of thickness because plasticizers are diminishing. All causes are documented [2-4] and confirmed by bend testing of material (film mPVC) in cold conditions, within which cracks are visually searched. An example is in the following paragraph.

Ranges of cracks in the surface of the test sample at various temperatures of test were determined during the testing. The results are summarized in Table 1.

Table 1. Bend testing of material at low temperatures.

The test sample	**Temperature** [°C]	**The upper surface – cross** [-]
1	23	0 - crack (slightly)
2	15	2 - larger cracks
3	5	3 - the sample cracked across the width or fragmented
4	0	3 - the sample cracked across the width or fragmented
5	-5	3 - the sample cracked across the width or fragmented

The performed measurements show that the waterproofing membrane does not perform its function after a very short time (approximately 6 years after incorporation into the construction in 2010). As it is apparent from the results of the bending of sample, the cracking occurs at temperatures around 15 °C. There is a strong cracking waterproof material over the entire width at low temperatures of about 0 °C. The condition of the waterproofing sheet of material can be documented by a variety of analytical methods such as IR, gas chromatography, and other.

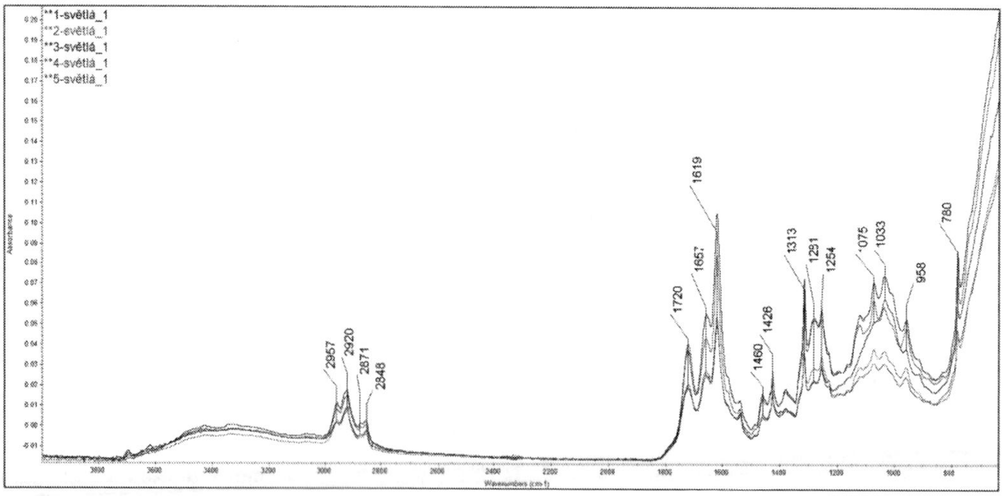

Fig. 1. Documentation of changes in mass of mPVC between covered and uncovered material.

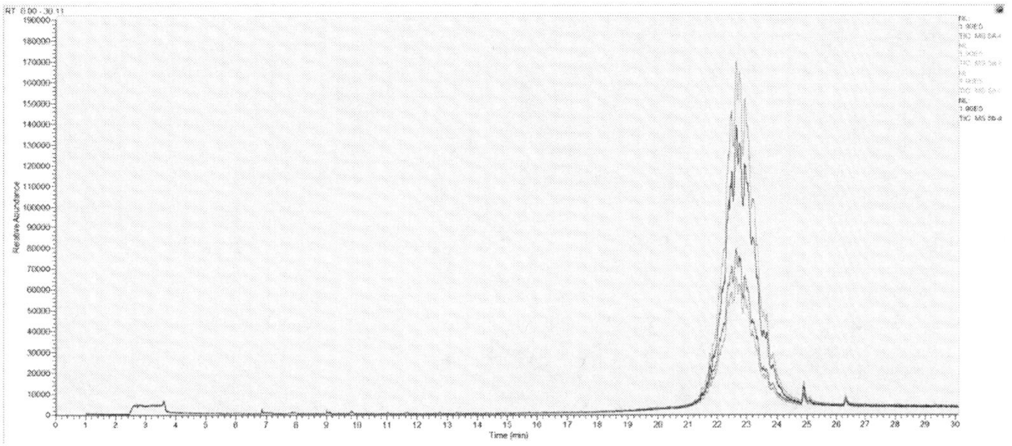

Fig. 2. Documentation of softeners and their comparison between covered and uncovered material.

Using these two methods, it is possible to obtain information on whether there were changes in the structure of the waterproofing material (IR analysis) and can be identified the pathological changes of the plasticizer. However, these processes cannot be precisely quantified. The using gas chromatography (English Gas chromatography, GC abbreviation) can cause loss of plasticizers and quantified.

These analytical methods in combination with the use of microscopic technology allow to identify defects and subsequently determined the remaining useful life, if there exists. It is possible to identify further possible manufacturing defects and their consequences apart from visual evidence of migration of plasticizers.

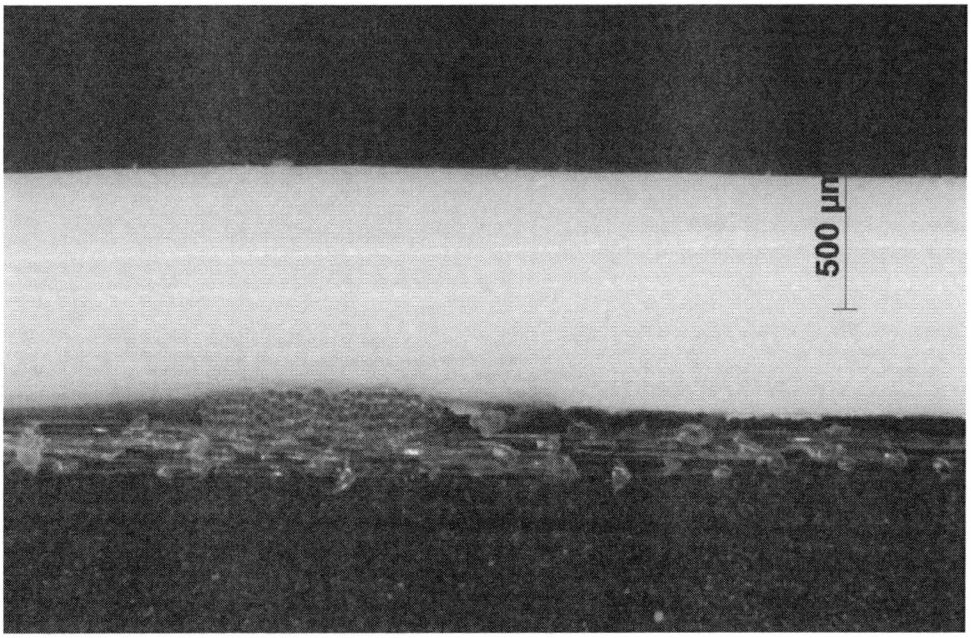

Fig. 3. The optimal condition of mPVC foil.

 DOI: 10.1051/matecconf/20179302008

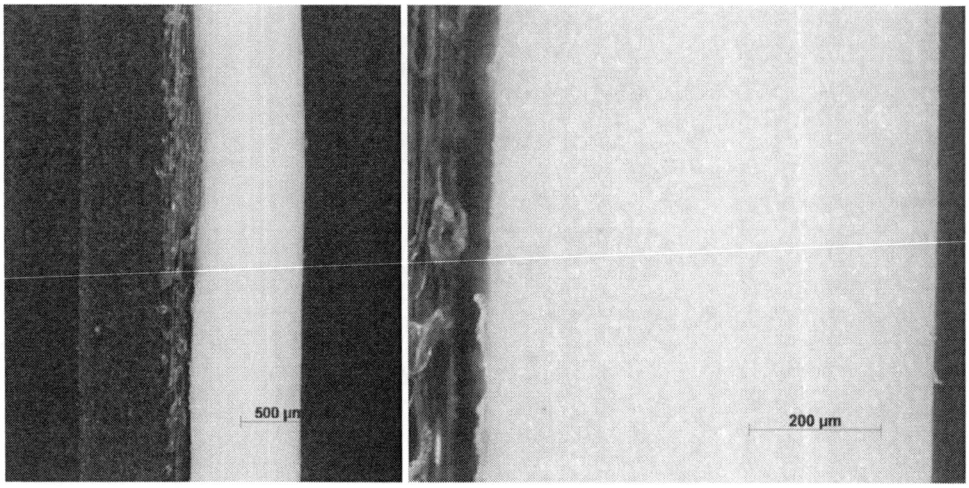

Fig. 4. Higher magnification shows inhomogeneity of the mPVC foil - The absorbency of humidity is the consequence of this inhomogeneity.

The following figures show the consequence.

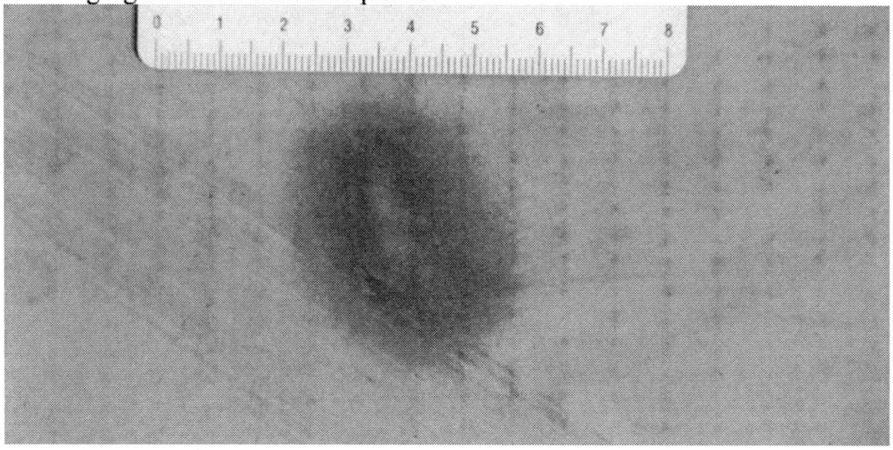

Fig. 5. Degradation spots on PVC foil.

Fig. 6. Degradation blisters on the foil mPVC.

DOI: 10.1051/matecconf/20179302008

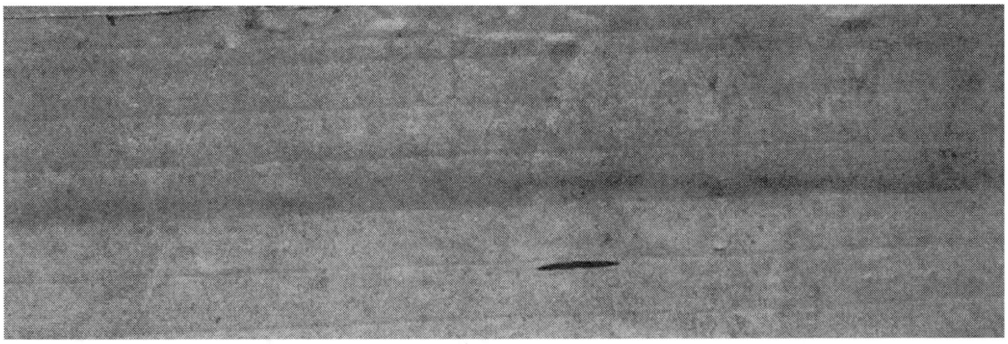

Fig. 7. The crack in mPVC foil.

Fig . 8. Notch, weakening on the top of the PVC foil. At the correct position of the reinforcement should be in the shape of the bottom (blue) oval.

 DOI: 10.1051/matecconf/20179302008

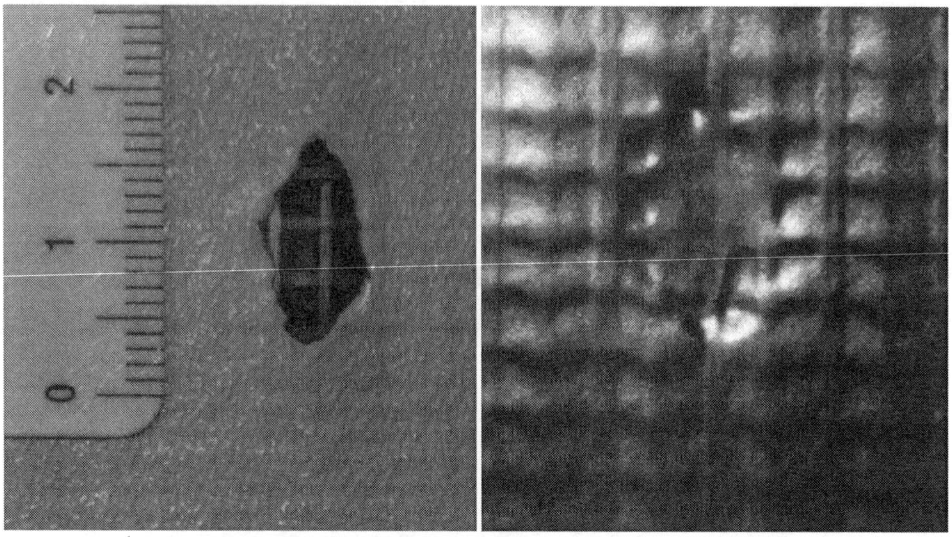

Fig. 9. Broken lower and upper surface of the mPVC foil.

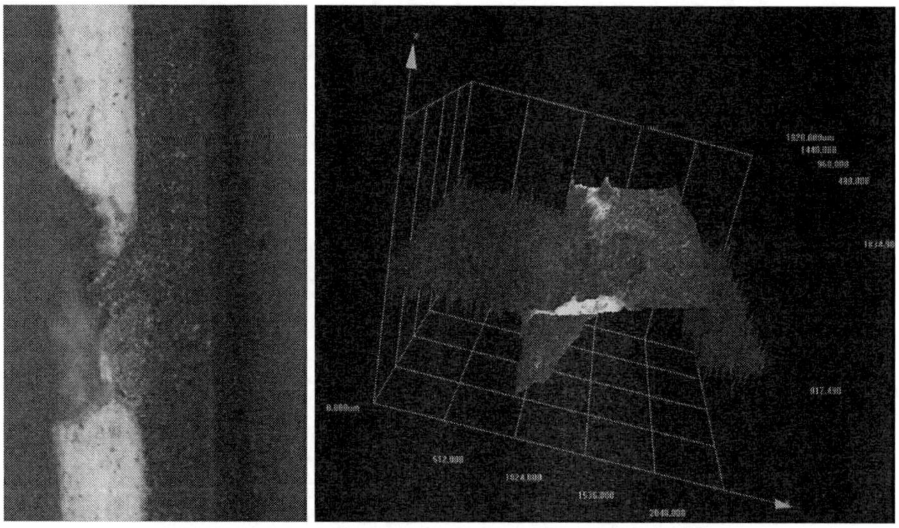

Fig. 10. Macro violations (left) and 3D modelling (right) of the violation by the mPVC foil.

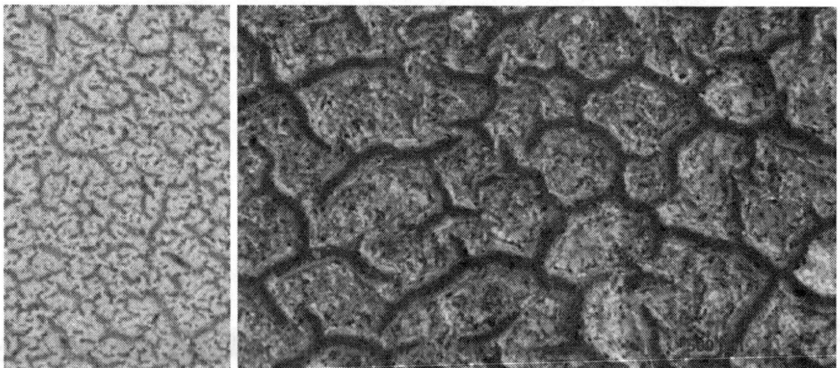

Fig. 11. Documentation "alligator ring" on the surface of the foil degrading mPVC and lower magnification (camera) on left side and micrograph of the surface on the right side.

 DOI: 10.1051/matecconf/20179302008

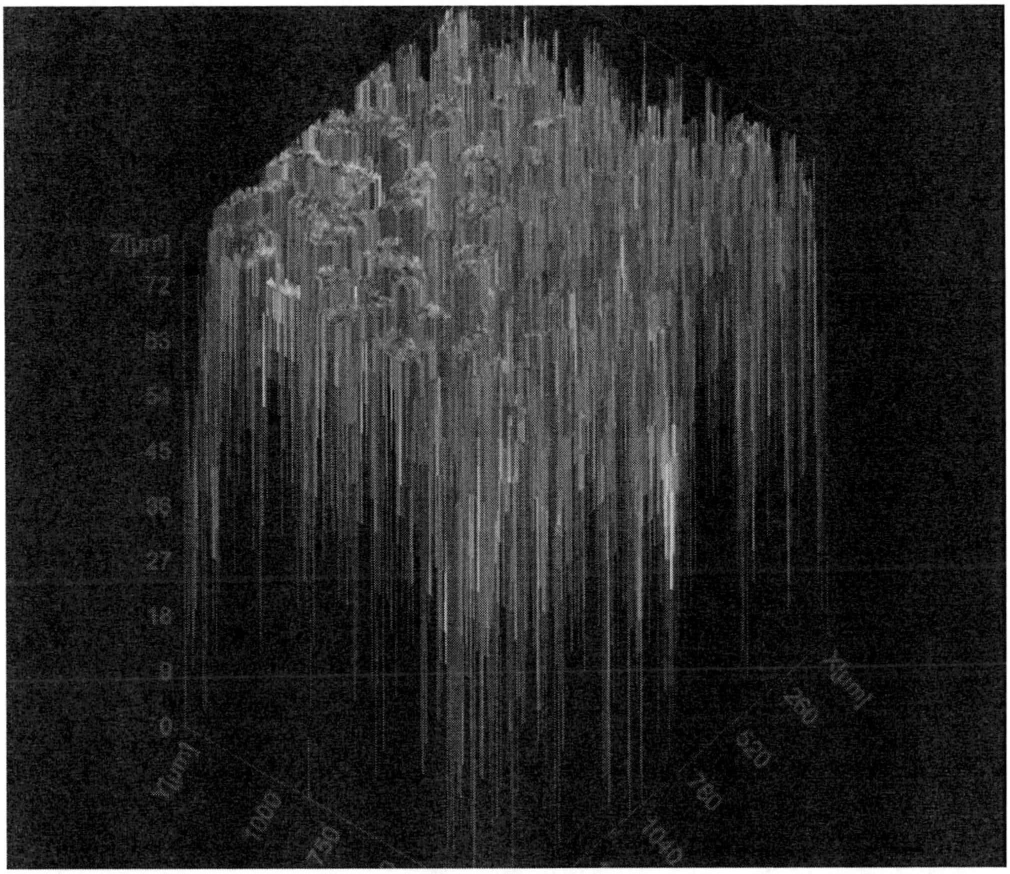

Fig. 12. Documentation of vaporizing surface of mPVC foil obtained from alligator ring, see fig. 11.

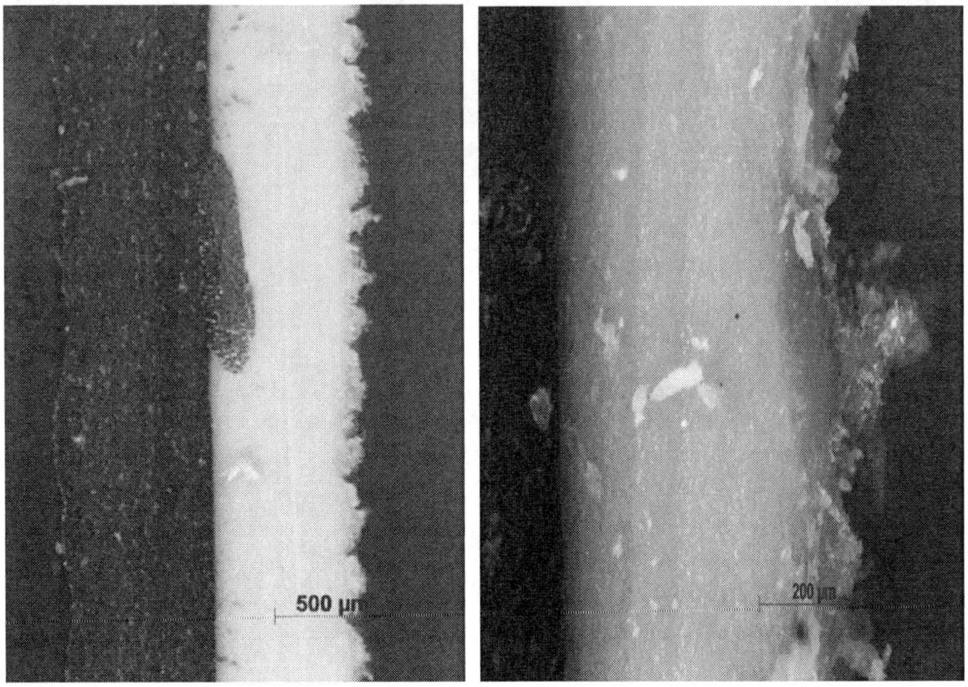

Fig. 13. Degradation of mPVC surface of the foil including a bad position of reinforcement.

Figure 13 illustrates degradation of the surface of the waterproofing membrane. It is a visible reduction in the volume of plasticizers. Figure 14 shows the state just before the collapse or collapse of the film.

The bursting of upper layers mPVC film is caused by forcefully stress. The overall loss of waterproof characteristics of waterproofing membrane is consequence of the rupture.

The figure 14 illustrates clearly visible crack in the top portion of the film that there are caused by the force stress of the film during its lifetime.

It is important to note that the total lifetime of the film is only 6 years old (Figure 1-4, 11-14). The durability is does not match the investor needs as well as the lifetime does not correspond with the EU standards and technical reports. According to the EOTA 002 (12/1999), the expected lifetime is at least 25 years.

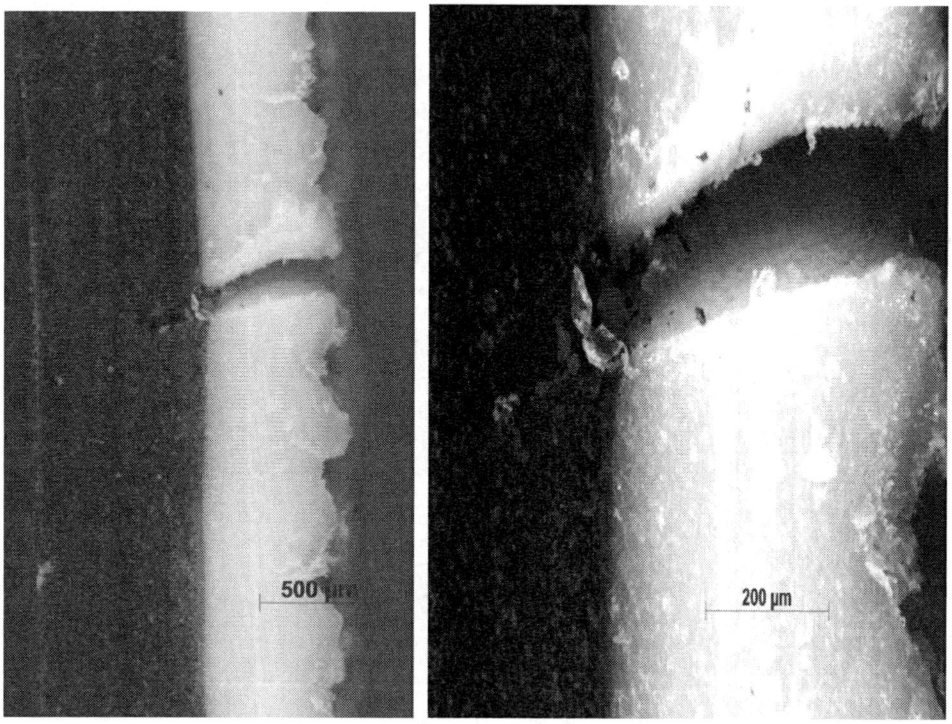

Fig. 14. Definitive end of functionality of PVC foil (waterproofing membrane) as a result of rupture.

3 Conclusion

The presented results are taken from executed expertise against specific manufacturers and specific material. It is necessary to say that the presented results are obtained by observing the cheapest type of mPVC foil on the Czech market. With respect to the occurring serious problems, it is recommended to stop the further production.

References

1. G.M. Vinhas, R.M. Souto-Maior, C.M. Lapa, Y.M. Bastos de Almeida, Mat. Res., **6** (4) (2003)
2. M. Novotný, Technical documentation, private database - expert witness (Unpublished)

3. FA CTU, Faculty of Engineering, Technical University of Liberec - technical documentation, test reports (Unpublished)
4. A.W.A.L., Technical documentation (Unpublished)

Indoor air quality – buildings design

Ingrid Juhásová Šenitková[1,*]

[1]Institute of Technology and Business in České Budějovice, Department of Civil Engineering, Okružní 517/10, 37001 České Budějovice Czech Republic

Abstract. Growing attention is being paid to indoor air quality as one of the main health and well-being factors. The indoor research is concerned mostly to indoor air chemicals within indoor engineering related to building design. The providing good indoor air quality can be achieved effectively by avoiding or reducing indoor air pollution sources and by selecting low-polluting building materials, both being low-cost and energy-efficient solutions. On the base of the last large experimental monitoring results, it was possible to know the level of selected indoor chemicals occurrence, rank them as well as to predict the tendencies of occurrence and establish the priorities for the future. There has been very limited attention to rigorous analysis of buildings actual environmental impacts to date. Healthy/green/sustainable building practices are typically applied in unsystematic and inconsistent ways often without resolution of inherent conflicts between and among such practices. Designers, products manufacturers, constructors, and owners declare their buildings and the applied technologies to be beneficial to the environment without validating those claims.

1 Introduction

Provided an environmental and economic will within our society will became aware of the real needs of solving the healthy/green/sustainable buildings problem – we are able to offer the models and strategies the tools, and criteria, and finally the construction technology and an architectural idiom to design, to build, and to maintain a sustainable building on short and long term as well as affordable. A health impact assessment and an environment impact assessment are needed to judge the real quality of a building and particularly indoor design.

As we spend most of our life indoors, it is easy to understand that the most important environment in relation to our health is the indoor environment. In the last several years, a growing body of scientific evidence has indicated that the air within buildings can be more seriously polluted than the outdoor air in even the largest and most industrialized cities. Indoor pollution sources that release gases or particles into the air are the primary cause of indoor air quality problems in the buildings [1].

Growing attention is being paid to indoor environment as one of the main health and well-being factors of building users. The main purpose of buildings is to create indoors more suitable for persons and processes than outdoors. However in buildings and shelters it

* Corresponding author: 14667@mail.vstecb.cz

is not only the thermal climate that is changed. The climate shell also stops the free air movement. The dilution of pollutants from close to man pollutants sources is diminished. Indoor air quality problems are more common in the buildings with insufficient ventilation. It is evident that there is need to increase the outdoor air supply in the indoor of buildings, to minimize the exposure of indoor pollutants. Furthermore, building materials, building constructions and indoor activities should be selected on the principle that the level of indoor air should be of the best quality or the concentration of negative agents should be as low as reasonably achievable. There are two classical approaches to reach a required level of indoor air quality in terms of pollutants. One is by dilution and other is by source control. Nowadays it seems that dilution is still the more accepted approach. It is the way of obtaining better indoor air quality [2].

This was and is the basis of the need for ventilation and for discussions on indoor air quality. Consequently, the main aim of building ventilation is to create the indoor air quality more suitable for persons and processes than what naturally occurs in the unventilated building, and to reintroduce the positive effect of being exposed to the wind, to dilute and remove the pollutants that man himself, his activities, and the indoor surrounding produce.

Sustainable buildings design depends significantly on the used criteria for the indoor environment and building design as well as for operation. Indoor environment also affects health, productivity and comfort of the occupants. Recent studies have shown that cost of the deteriorated indoor environment for the society, employer and building owner are often higher than the cost of energy used in the same buildings. An energy declaration without a declaration related to the indoor environment makes no sense. There is therefore a need for specifying criteria for the indoor environment for design, energy calculations, performance and operation.

The additional objective of the last research work is perceived air quality. The interaction between building materials, structures, HVAC systems and occupancy performance is continously observed [3].

2 Indoor environments

Environmental quality of indoors is the conceptual system principle of architectural, constructional and environmental design development. Healthy/green/sustainable buildings are determined with mass and environs aspect. Mass aspect is determined by the system Building-Material-Subject and environs aspect is influenced by Building-Fluid-Subject system. Indoor environmental factors are determined by operative perceived comfort and hygiene criteria of indoor physical, chemical and biological compounds. The proper healthy/green/sustainable building design is doubtless task of mass and energy flow reduction. The mass flow reductions mostly consist in material recovery [4,5]. The problems with indoor air quality occur when energy flow reduction works together with changed material base composition and used services technology became dominant. There is no contradiction between indoor air quality and energy performance, but the more mass is changed the more energy flow is needed.

The investigation of selected chemicals (volatile organic compounds, particles, odors, formaldehyde, radon and nitrogen oxides) as well as were the main subjects of the last period interest [1,6,7]. The systematic indoor research studies in various types of buildings were carried out within the national research projects. On the base of the last large experimental monitoring results, it was possible to know the level of selected indoor chemicals occurrence, rank them as well as to predict the tendencies of occurrence and establish the priorities for the future (Table 1).

Table 1. Indoor chemical pollutants occurrence ranking.

Pollutants	Occurrence [%]	Tendency
VOCs	85	↑
Odors	78	↑
PM_{10}	61	↑
Radon	35	↓
NO_2	19	↓

It has become evident that variety building materials are the major source of indoor volatile organic compounds. The resulting interactions between indoor materials and indoor pollutants may significantly alter the concentrations of contaminants in indoor air. Interior surfaces are generally accepted as the main source of VOCs (Volatile Organic Compounds) emissions [1]. There has been growing evidence that building materials can affect the transport and exposure of indoor VOCs by sorption. The re-emission of adsorbed VOCs can dramatically elevate VOC concentrations in the indoor environment for months or years after a source event [1,2]. The surface sorption of materials emissions can reduce the concentration and improve the perceived air quality. Therefore, interaction between individual indoor materials (sorption ability of materials) is appropriate to study.

Presented results confirmed the highest priority of chemicals especially volatile organic compounds, odors, particles, radon and some oxidants. Indoor air quality become as integrating factor for perceived comfort and hygiene of indoors. Some interesting prodigies meet each other as environment, materials, constructions, HVAC systems, health, economy and productivity. The providing good indoor air quality can be achieved effectively by avoiding or reducing indoor air pollution sources and by selecting low-polluting building materials, both being low-cost and energy-efficient solutions.

Buildings may cause illness to their occupants, producing symptoms like irritation in the nose, throat, eyes, and skin, as well as shortness of breath, nausea, dizziness, and fatigue. These symptoms are commonly referred to as the Sick Building Syndrome (SBS). Such illness is usually minor and disappears when the sufferer leaves the building. Low rates of ventilation are often mentioned as a contributory factor. Allergy to indoor air contaminants is the main factor in sick building syndrome. These contaminants can be in the form of fine dust, fumes, fibers, odors or gases which are not removed by air conditioning filters or cleaning. Health effects from indoor air pollutants may be experienced soon after exposure or, possibly, years later. Immediate effects may show up after a single exposure or repeated exposures. These include irritation of the eyes, nose, and throat, headaches, dizziness, and fatigue. Such immediate effects are usually short-term and treatable. Sometimes the treatment is simply eliminating the person's exposure to the source of the pollution, if it can be identified. Symptoms of some diseases, including asthma, hypersensitivity pneumonitis, and humidifier fever, may also show up soon after exposure to some indoor air pollutants. The likelihood of immediate reactions to indoor air pollutants depends on several factors. Age and preexisting medical conditions are two important influences. In other cases, whether a person reacts to a pollutant depends on individual sensitivity, which varies tremendously from person to person. Some people can become sensitized to biological pollutants after repeated exposures, and it appears that some people can become sensitized to chemical pollutants as well. Another way to judge whether living environment has or could develop indoor air problems is to identify potential sources of indoor air pollution. Although the presence of such sources does not necessarily mean that person have an indoor air quality problem, being aware of the type and number of potential sources is an important step toward assessing the air quality in the indoor environment.

3 Buildings design

When designing a building there are mainly four different factors to consider: regulations, customer requirements, knowledge and material information. These four factors have different goals, but one of them is to create a healthy buildings. Unfortunately there are also other goals that have to be fulfilled for each different factor, e.g. low cost, much space. The different goals are not always mutual for the different factors and that makes it sometimes hard to find good solutions for the indoor environment even when enough knowledge is available.

There can be little doubt that enormous progress has been made in improving the understanding of the design and construction practices that produce healthy buildings in terms of the impacts of buildings on occupant`s health and well-being. A broader definition of healthy buildings included not only the impacts of buildings on their occupants but also on the larger environment. A healthy building is one that adversely affects neither the health of its occupants nor the larger environment.

The guidelines for healthy buildings project consists of design recommendations, design criteria and parameters, methods of suitable solution design and references to standard control. The building design should be based on a determination of what is important for environmental safety and established priorities for design efforts. The building design should include assessment of the amounts of surface area and mass of major building materials and products that will be used. The overview of the building use and the appropriate ventilation strategy as well as the detailed plan for the operation and maintenance of the building over its entire life cycle should be included in building design process. Analyzed specific features of buildings with sufficient indoor air quality to achieve required characteristic of healthy building, four main groups of tools do exist:

- Architectural and essential technical concept,
- Detailed architectural and construction design,
- Building environmental engineering,
- Operating and controlling systems.

The recognition of local and global environmental problems have stimulated considerable process which has been made during the past thirty years toward a more complete understanding of design and construction requirements for healthy buildings. The building design and construction toward sustainable practices were slowly shifted by awareness of environmental problems. Many environmental-friendly design and construction principles are widely accepted, but designers and constructors often fail to recommended practices or their clients fail to adopt state-of-the-art practices. Reasons for these failures include preference for and adherence to traditional practices, lack of expertise or short-term financial considerations. Meanwhile, awareness of environmental problems has produced an emerging shift toward purportedly green building practices that presume to include buildings that are healthy for their occupants.

4 The differences between green and sustainable

The idea of energy efficient, healthy buildings has been around for a long time, so why is it just now that the concept of green or sustainable building is entering the mainstream and catching the attention. There are probably several reasons; global warming, rising energy costs, the growing awareness and liability costs associated with sick building syndrome, declining oil reserves, and concerns about our limited water supply. The list goes on, but whatever the reason or reasons, sustainable building is a concept whose time has come. However, the words green and sustainable are often used interchangeably, and sustainable has a more precise meaning that is often obscured, distorted, and diluted by the commercialization and marketing of the green movement.

My personal definition of green is relatively simple. A building's design is green if its serves to reduce many of the harmful impacts buildings have on environment and building's inhabitants. So green building design revolves around four key issues:

- Designing for energy efficiency including the use of renewal energy sources,
- Creating a healthy indoor air environment with adequate ventilation,
- Making material choices that minimize volatile organic compounds,
- Specifying building materials and resources that are sustainable, have low embodied energy, and produce a minimal amount of upstream environmental impact.

Sustainability is ultimately a question of mass flows. Thus, the best way to make buildings more sustainable is to reduce mass flows. There are four main strategies for reducing mass flows in buildings:

- Re-use existing buildings,
- Re-use building materials,
- Recycle building materials that cannot be reused,
- Accommodate the same function in buildings.

In broadest sense, sustainability is an ethic which seeks to minimize the impact of human actions on the environment, and which strives to meet the needs of the present without compromising the ability of future generations to meet their own. Sustainable planning, design and development, then refers to strategies of practice and intervention aimed to accommodate present human needs in a manner which considers the full spectrum of natural and cultural system native to a site, and seeks a course of action minimally disruptive to those systems and which, ideally, results in their enhancement and long term viability. Sustainable design emphasizes relationships and processes rather than objects.

5 Discussions

While the designers maybe are motivated to create sustainable buildings, they are faced with number daunting challenges when they begin an actual project. First sustainable buildings have many definitions it is difficult to consistently identify what strategies should be included with sustainable project. A considerable part of this problem is that information is developing and changing rapidly. Second, there is an overwhelming amount of information on the subject that is either too time consuming to sift through or too fragmented to understand as a coherent whole. Third, while there already excellent sustainable building rating systems are available, they focus more on the strategies and less on the process required to make them work on an actual project.

There has been very limited attention to rigorous analysis of buildings actual environmental impacts to date. Healthy/green/sustainable building practices are typically applied in unsystematic and inconsistent ways often without resolution of inherent conflicts between and among such practices. Designers, products manufacturers, constructors, and owners declare their buildings and the applied technologies to be beneficial to the environment without validating these claims. Those seeking guidance to construct green buildings accept these claims without further analysis. This has resulted in the establishment and increasingly widespread acceptance of many practices that are of undetermined environmental impact. It has also produced a number of guidance documents that have begun to codify these practices. Much more work is required to determine the total environmental impacts of the buildings that are built. This work should include detailed evaluation of the pollution emissions, operation and use of buildings, and embodied and life cycle resource consumption.

The design healthy/green/sustainable buildings should be based on a determination of what is important as well as on establishing of design efforts priorities. The design should include the identification of major materials and products that will be used, an overview of use of the building and the appropriate ventilation strategy and detailed plan for the operation and maintenance of the building over its entire life cycle.

Each strategy requires actions to be taken in some or of a building's life cycle, which is organized into these parts: planning, design, construction occupancy, and next use. The planning phase includes project initiation, programming and site selection. The schematic design, design development and construction documents are subparts of design process. The planning, design and construction parts represent building delivery process. The sustainable building guide must also include the entire occupancy period of the facility, particularly since public agency typically owns and operates these buildings. During occupancy many of the strategies must be implemented to be effective and the environmental impacts of operating and maintenance part of building life cycle offers critical opportunities to optimize the energy performance of a facility. At the end of the operating life of a building for a particular function, the next use of the facility will be analyzed and the process of planning, design construction and operation will begin once again.

6 Conclusions

The building design has to respect qualitative and quantitative requirements of indoor air quality parameters for environmental safety from space and time differences point of view. As the first stage of building design is the most important for final indoor air quality more and more by clear constructional architecture the benefit of environmental safety, good indoor air quality for health and productivity are dominant. The building design concerning to heath of occupants and hygienic performance regime can be realized only by interdisciplinary team of professionals respecting the healthy/green/sustainable building and indoor environment design.

Sustainable indoor environment design it is not easy task especially for designers. They do not practicing systematic attention to architectural, constructional and environmental attention together. Although they have some ethic motivation in using the newest indoor sciences knowledge, they meet a lot of problems in the real design process usually. A bulk of strategies do exist in addition they are permanently changed they are time-demanding and not quite understandable.

The author is grateful to national grant agency for supporting the indoor air sciences research project.

References

1. I. Senitkova, T. Tomcik, Advanced Science Letters, **19**, 9 (2013)
2. I. Senitkova, M. Bucakova, Selected Scientific Papers, **3**, 45-56 (2008)
3. I. Senitkova, Material Science Engineering C, **36**, 6 (2014)
4. M. Culakova, I.Senitkova, A. Paulikova, Pollack Periodika, **7**, 33-44 (2012)
5. I. Senitkova, P. Bednarova, JP Journal of Heat and Mass Transfer, **11**, 29-42 (2015)
6. I. Senitkova, JP Journal of Heat and Mass Transfer, **12**, 89-102 (2015)
7. I. Juhasova Senitkova, JP Journal of Heat and Mass Transfer, **13**, 11 (2016)

 DOI: 10.1051/matecconf/20179303002

Problems of heating and ventilation in small university lecture hall

Mária Budiaková[1,*]

[1]Slovak University of Technology in Bratislava, Nám. slobody 19, 812 45 Bratislava, Slovakia

Abstract. The paper is oriented on the problems of heating and ventilation in small university lecture hall. Providing the optimal parameters of thermal comfort is immensely important for students of university in small university lecture hall. Incorrectly functioning heating and ventilation can disrupt thermal comfort. Reconstruction took place in small university lecture hall. The original windows were exchanged for modern wood tight windows. Experimental measurements were carried out in winter season in 2016 in this small university lecture hall, in order to evaluate the thermal comfort and CO_2 concentration after the reconstruction. The device Testo 480 was used for measurements. Obtained values of air temperature, air relative humidity, air velocity, globe temperature, indexes PMV, PPD and CO_2 concentration are presented in the graphs. Heating, ventilation and operation of small university lecture hall were evaluated on the basis of parameters of thermal comfort and CO_2 concentration. Principles how to design the heating and ventilation in small university lecture halls are in the conclusion of the paper. Furthermore, there are presented recommendations how to operate the heating and ventilation in existing small university lecture halls.

1 Introduction

In old university buildings, it is very difficult rather impossible to provide thermal comfort, what also relates to incorrect architectural design, old not tight windows, not insulated peripheral wall, insufficient controlled and regulated heating system and to missing mechanical ventilation system. Therefore, new university buildings should have such thermotechnical parameters, energy economical regulated heating system and mechanical ventilation system that would enable provision of thermal comfort in small university lecture halls with high quality.

Securing of thermal comfort in small university lecture hall is very important because students spend in schoolrooms the majority of their time in school. Thermal comfort in the small university lecture hall is defined as the state of mind that expresses satisfaction with the surrounding environment. The fundamental quantities for the evaluation of the thermal comfort are internal air temperature, operative temperature, globe temperature, air relative humidity and air velocity [1]. Then the thermal comfort is evaluated with index PMV (Predicted mean vote) and index PPD (Predicted percentage dissatisfied) [2-3]. Not

* Corresponding author: budiakova@fa.stuba.sk

fulfilling the parameters of the thermal comfort in the small university lecture hall contributes to the high sickness rate of students especially in the winter [4-5]. Not correctly provided thermal comfort also increases energy intensity of university schoolrooms.

Classroom environment is characterized by a high density of students and so the main problem is usually indoor air quality especially high CO_2 concentration, which is generated by human presence (metabolism, respiration). CO_2 levels that are unusually high indoor climate may cause students drowsiness, headaches, or decrease of their activity. Humans are the main indoor source of carbon dioxide in the most of the buildings. Indoor CO_2 levels are an indicator of the adequacy of outdoor air ventilation relative to indoor student density and metabolic activity. In order to eliminate the most complaints, the total indoor CO_2 level should be reduced to a difference of less than 600 ppm above outdoor levels.

In the small university lecture hall, the minimal value of an operative temperature 20°C must be fulfilled. Small university lecture halls in Slovakia are heated by convective heating elements, Fig. 1. Big massive sills and decoration overlays of convective heating elements significantly increase energy demands and menace the provision of thermal comfort, therefore their application is undesirable. The right size and position of convective heating elements must ensure that the local thermal discomfort will not arise. Local thermal discomfort (the thermal dissatisfaction) can also be caused by unwanted cooling or heating of one particular part of the body [6]. The most common cause of the local thermal discomfort is the draught but local discomfort can also be caused by an abnormally high vertical temperature difference between head and ankles, by too warm or too cool floor, or by too high radiant temperature asymmetry. People are most sensitive to radiant asymmetry caused by a warm ceiling, a cool wall (windows, glazed facade), a cool ceiling or by a warm wall. The major problem in the small university lecture hall are cool windows; therefore, their surface should be minimized to just to provide the daily lighting.

Thermal comfort in the small university lecture halls in Slovakia is not provided and often arises the local thermal discomfort. Therefore, this research was focused on the evaluation of the thermal comfort parameters and CO_2 concentration in small lecture hall.

Fig. 1. Small university lecture hall with convective heating elements.

2 Characteristic of experimental measurements

Experimental measurements were carried out in the small university lecture hall - Fig. 1 at the Slovak University of Technology in Bratislava, Faculty of Architecture in the beginning of March 2016. The reconstruction of this lecture hall took place before. New wooden tight windows were installed.

The aim of the measurements was to record the parameters of the thermal comfort: air temperature, air relative humidity, air velocity, globe temperature, index PMV and index PPD; and the parameters of CO_2 concentration.

The measurements were carried out in the small university lecture hall with the sizes 6,6 x 14,5 m in the center of the room in height level 1,10 m. There are five windows built into the external wall with the sizes 2,3 x 2,3 m. Under each window is recurrent convective heating element. Continuous massive sills are above the heating elements. Four convective heating elements are covered by the decorative slats.

The parameters of the thermal comfort and CO_2 concentration were recorded with the device Testo 480. Input data in the measurements were: metabolic rate 1,0 met, clothing insulate 1,0 clo. The measurements of the thermal comfort were carried out in three successive phases that were three tutorial lessons. Each phase took 60 minutes; last 7-10 minutes was break. Different group of students took part in each phase: I – Fig. 2, II – Fig. 3, III – Fig. 4. Students exchanged during the break. The doors and all the small windows were also opened during the break to provide the natural ventilation. Measurement of CO_2 concentration was carried out in the following week in one phase, Fig. 5.

Outdoor air temperature was measured and recorded by separate device. Outdoor air temperature increased during the measurement of thermal comfort from the value 10,3 °C to the value 11,1 °C. Outdoor air temperature increased during the measurement of CO_2 concentration from the value 9,1 °C to the value 10,3 °C.

Fig. 2. Measurement in the first phase.

Fig. 3. Measurement in the second phase.

Fig. 4. Measurement in the third phase.

Fig. 5. Measurement of CO_2 concentration.

3 Results and analysis of experimental measurements

The values of air temperature are showed in Fig. 6 in all three phases in the height level 1,1 in the centre of the room. The values of air relative humidity are showed in Fig. 7, the values of air velocity are showed in Fig. 8, the values of index PMV (Predicted mean vote) are showed in Fig. 9, the values of index PPD (Predicted percentage dissatisfied) are showed in Fig. 10.

Fig. 10 shows the values of CO_2 concentration in the first phase during 60 minutes in the height of 1,1 m above the floor level in the center of the room.

Air temperature (Fig. 6) in the small university lecture hall was high, what relates to more causes. There was no regulation of the heating system after the reconstruction, and after the installation of new windows. Heat losses decreased in the new windows, which was not reflected in the operation of convective heating elements. The low air temperatures in the beginning and in the end of tutorial session show the natural ventilation by windows and door. This natural ventilation during the 7-10 minute break is not sufficient, not only in terms of CO_2 load for students, but also in terms of the amount of heat transmitted by students in the interior. This problem shows constantly increasing air temperature in the first, then in the second and the biggest values of air temperature were in the third phase of the measurements.

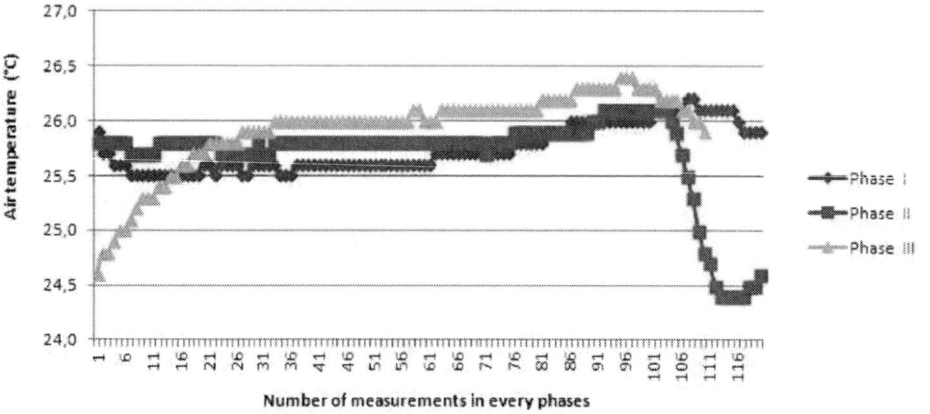

Fig. 6. Values of air temperature in all three phases.

Air relative humidity (Fig. 7) in the small university lecture hall does not have significantly critical values. It is obvious from the graph that air relative humidity values were optimizing itself during the break when there was ventilation. These optimal values kept only couple of minutes after the break. The problems of indoor climate in the lecture hall are pointed out by increasing values of air relative humidity during the first phase, following-up the second and the biggest values of air relative humidity were in the third phase of the measurements.

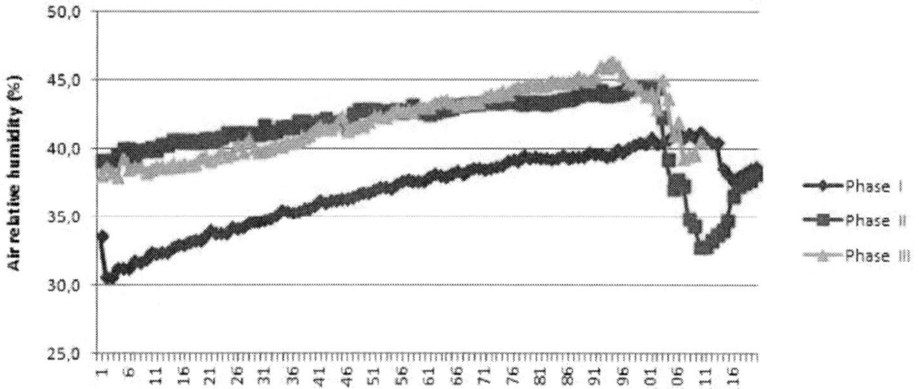

Fig. 7. Values of air relative humidity in all three phases.

Air velocity (Fig. 8) in the small university lecture hall had appropriate values during the tutorial session. High values of air velocity were measured during the break, when there was ventilation and the draught arose between the windows and the door.

Measured values of Index PMV (Fig. 9) and Index PPD (Fig. 10) points out on serious problems with the thermal comfort. Optimal values were not achieved during the tutorial sessions, only short-time during the natural ventilation in breaks. Index PMV and PPD value had still tolerable value during the phase I and II. During the last phase III, Index PMV and PPD value did not have tolerable value in the second half of the tutorial session. Successive increasing values of Index PMV and PPD value during all the phases point out on the problems with the heating system and the missing mechanical ventilation system.

 DOI: 10.1051/matecconf/20179303002

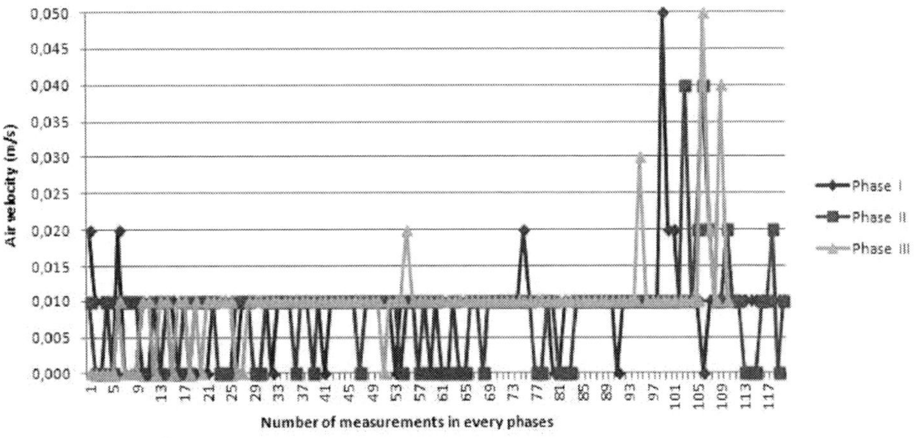

Fig. 8. Values of air velocity in all three phases.

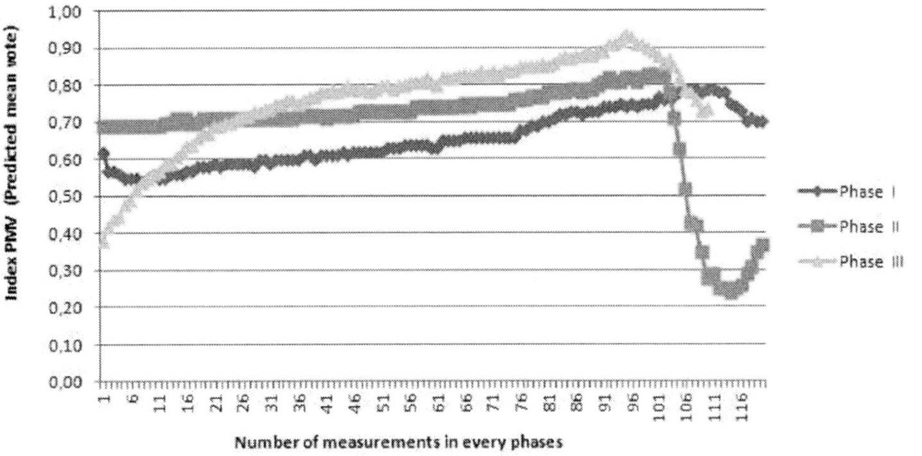

Fig. 9. Values of index PMV in all three phases.

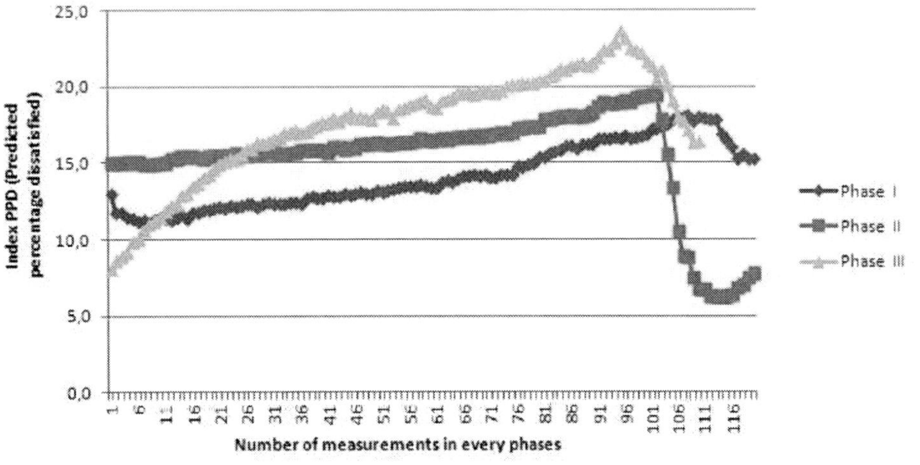

Fig. 10. Values of index PPD in all three phases.

 DOI: 10.1051/matecconf/20179303002

The values of CO_2 concentration (Fig. 11) linearly increased from the value 1327 ppm to the value 2739 ppm. These values are inadmissible for the work of students in the small university lecture hall. During the small break between lessons, the doors and all the small windows were also opened. This natural ventilation contributed to the decrease of CO_2 concentration on the value 2530 ppm, which might be considered as very inefficient. The students complained the most about the quality of the air in the small university lecture hall, had problems with breathing and they had minimal concentration. Results of these measurements showed that during the break, the natural ventilation by doors is completely insufficient. Exchange of the windows brought the quality into the reconstruction, that has not been finished because of the shortage of money and the forced ventilation has not been built. Students' complaints during the lesson were significant, therefore, not only small parts of windows must be opened but also big parts of windows. This caused big paradox; tight energy saving windows are causing enormous heat losses by permanently opened leaves of windows and thus the energy demands have grown. In order to increase the air quality and to achieve energy savings in schoolrooms, it is inevitable to build forced ventilation system.

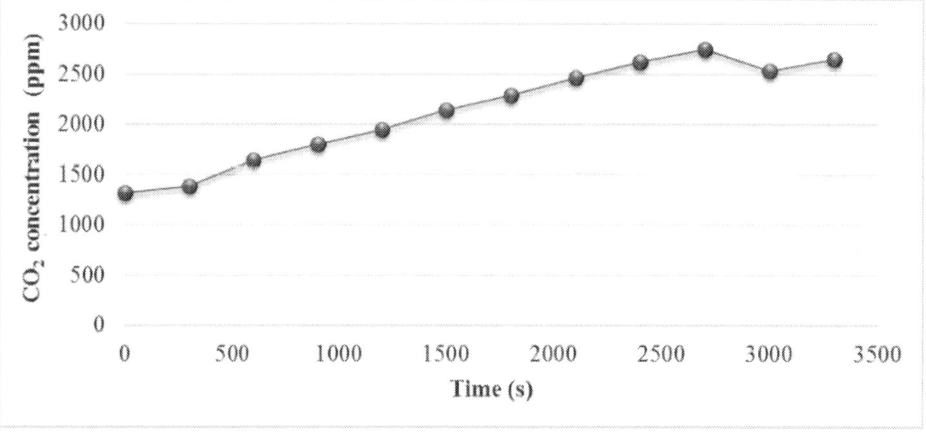

Fig. 11. Values of CO_2 concentration.

4 Conclusions

Experimental measurements in the small university lecture hall showed the problems with the provision of the thermal comfort. The reconstruction was not thorough. Windows were exchanged for better quality windows, but the massive sill and decorative overlay of convective heating elements were not removed. There was no regulation of the heating system after the reconstruction and the installation of new windows. This fact added to the problems in operation of heating system. There is no controlling element of the heating system, which would adjust the power of convective heating elements to current conditions, number of the students in the small university lecture hall.

Experimental measurements showed that natural ventilation by doors and windows during the break is completely insufficient. Big problems arose with the provision of the air quality because the values of CO_2 concentration reached inadmissible values. The students complained the air quality in the small university lecture hall, had big difficulties with breathing and their concentration on the subject decreased sharply. Exchange of windows brought quality into the reconstruction, but represents just the first step; the next step must be the forced ventilation. Students' complaints during the lesson were significant, therefore, the windows must be opened during the entire lesson. It causes big heat losses and significant increase of energy consumption. It is inevitable to build forced ventilation system in order to increase the air quality and to achieve energy savings in the schoolrooms. Total reconstruction can bring meaningful results.

During the design of new small university lecture hall, it is necessary to take into the account the requirements for providing the thermal comfort. Therefore, the architectural design, the design of the heating system and the design of the forced ventilation system or the air-conditioning system must be focused on the provision of the optimal indoor climate in a lecture hall.

Scientific paper is published with the support of grant agency KEGA, grant n. 039STU-4/2014.

References

1. L. Bánhidi, L. Kajtár, *Komfortelmélet (Comfort Theory)* (Muegyetemi kiadó, Budapest, 2000)
2. *STN EN ISO 7730, Ergonomics of the thermal environment - Analytical determination and interpretation of thermal comfort using calculation of the PMV and PPD indices and local thermal comfort criteria* (2006)
3. *STN EN ISO 7726, Ergonomics of the thermal environment. Instruments for measuring physical quantities* (2003)
4. D. Etheridge, *Natural Ventilation of Buildings* (John Wiley & Sons, United Kingdom, 2011)
5. M. Jokl, *Zdravé obytné a pracovní prostředí* (Academia, Praha, 2002)
6. *STN EN 15251, Indoor environmental input parameters for design and assessment of energy performance of buildings addressing indoor air quality, thermal environment, lighting and acoustics* (2008)

MATEC Web of Conferences **93**, 03003 (2017) DOI: 10.1051/matecconf/20179303003
Building Defects 2016

The influence of ventilated façade on sound insulation properties of envelope walls

Zuzana Fišarová[1], *Lubor* Kalousek[1,*], *Michal* Frank[2], and *Roman* Brzoň[1]

[1]Brno University of Technology, Faculty of Civil Engineering, Veveří 95, 602 00 Brno, Czech Republic
[2]Research Institute for Building Materials, Hněvkovského 30/65, 617 00 Brno, Czech Republic

Abstract. Presented article deals with sound insulation properties of timber structures' envelope walls. Particularly, the influence of heavy board ventilated façade on laboratory airborne sound insulation R and R_w in dB was studied. The installation method and gaps between façade boards can cause building defects originating in overrating the influence of ventilated cladding on envelope wall acoustic parameters. Real constructions were built for the experimental purposes and measurements, one with gaps between boards and one with simply eliminated gaps for mutual comparison. The results obtained were processed to make tables and graphs and to derive recommendations for the design of this type of constructions involving the general installation method of façade boards. Detailed results are depicted in conclusions.

1 Introduction

Due to growing rate of timber structure buildings in the Czech market, the issue of protecting the inner environment and its inhabitants against excessive noise slowly emerges. The issue of insufficient sound insulation of the envelope wall was frequently neglected especially in small buildings for housing. A significant change came with amendment of Act No. 258/2000 Coll., on public health protection, and following amendment of some related laws. This amendment reflects particularly increasing requirements on protection against noise adverse effects and in this respect fundamentally toughens the whole building approval process. The amendment is applied on residential building as well as on educational, healthcare and similar buildings.

The presented experiment was conducted with timber structures' envelope wall. However, the findings on the influence of the ventilated façade are generally valid also for other types of walls, e.g. brick walls or aerated concrete block walls.

* Corresponding author: kalousek.l@fce.vutbr.cz

2 General requirements on sound insulation of envelope walls

General requirements on sound insulation of envelope walls and their parts are given in ČSN 73 0532/2010 – Acoustics – Protection against noise in buildings and evaluation of acoustic properties of building elements – Requirements, [1].

Whether a real finished construction meets the requirements is verified by an experiment on building site according to ČSN EN ISO 140-5, [2].

In designing phase, engineers rely on calculations and values given by producers or suppliers of the envelope construction or its parts.

3 Sound insulation defects of envelope walls

As mentioned in the introduction, errors can occur in design and execution of the envelope walls. One of the most significant errors is underestimating the influence of specific insulation material on the sound insulation properties of the resulting structure. This topic was by authors described in detail in [3], where the comparison of acoustic properties of envelope walls before and after ETICS application is the main topic. Among other a significant decrease of sound reduction index R from approx. 180 to 600 Hz due to some insulation materials was analyzed in this publication.

Another frequently occurring phenomenon is the absence of plaster on the outside of envelope wall due to presence of thermal insulation layer. Designers mistakenly think that a brick or a block has the same sound insulation properties with one sided plaster and with both sided plaster.

Last but not least mistake is overestimating the influence of outer ventilated cladding which can easily occur especially with heavy facade boards. This issue is the main topic of this article and currently available data are widened by measurements on real constructions.

4 Experiment description

The prime aim of the experiment was a general widening of knowledge on sound insulation properties of building envelope construction made of D3D system according to utility model No. 15073 U1, [4]. Mainly the influence of different contact thermal insulation materials was evaluated as described in [5] and followingly, the benefit of heavy facade boards was analyzed. Through the whole project, numerous structures were examined. All laboratory sound insulation measurements were conducted according to valid methodics in acoustic echo chamber of the Institute of Physical Engineering, laboratory accredited by ČIA under No. 1016, Brno University of Technology.

5 Specification of measured structures

Three basic groups of tests were conducted gradually during the construction. The first group is represented by measurements of basic D3D construction insulated by 2 x 100 mm Steico Flex and covered by Egger boards and with external ventilated cladding made of heavy CEMVIN facade boards. For detailed description see Fig. 1.

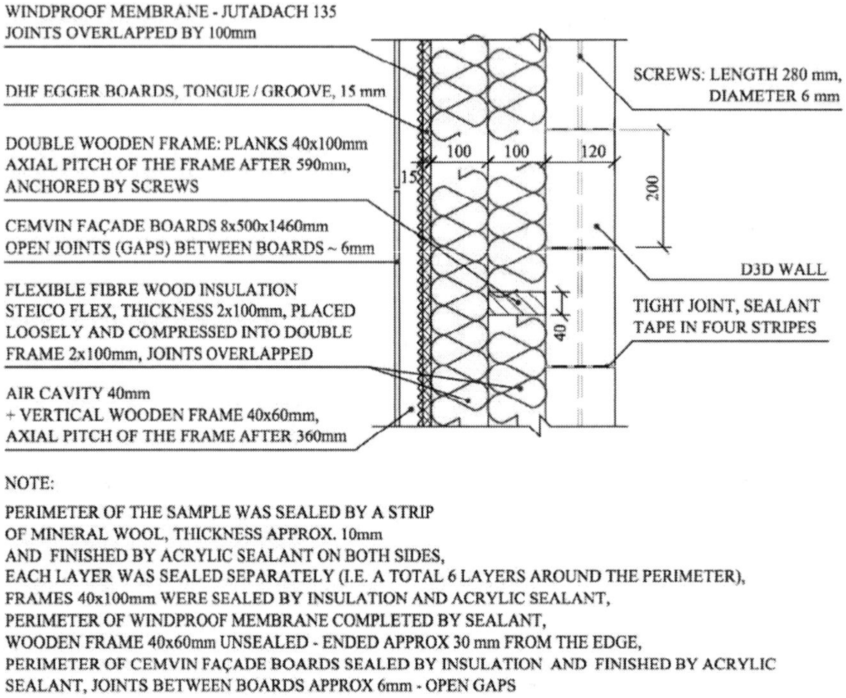

Fig. 1. First measurement group – basic structure with Steico Flex thermal insulation and external ventilated cladding.

For the second group of measurements, Steico Flex insulation was replaced by ISOVER UNI, all other parts remained unchanged (see Fig. 2).

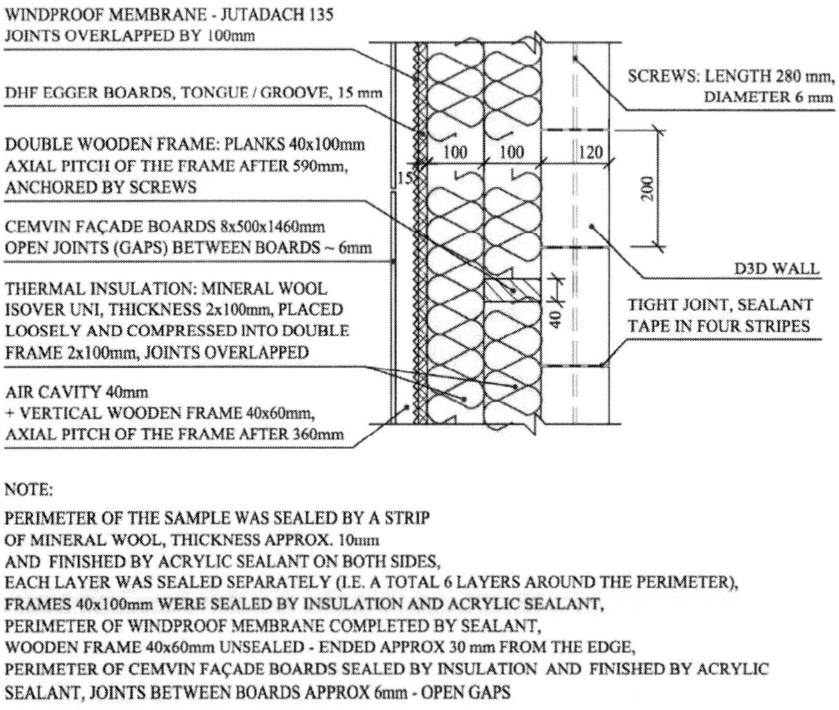

Fig. 2. Second group of measurements – basic structure with ISOVER UNI thermal insulation and external ventilated cladding.

 DOI: 10.1051/matecconf/20179303003

For the third group of measurements, a sample with simple external cladding gap elimination was constructed and the influence on sound insulation was examined. The gaps, i.e. free joints between façade boards, on existing structure measured in previous step were precisely sealed and covered with an aluminium tape as seen in Fig. 3 and 4. The other parts of previous composition remained unchanged.

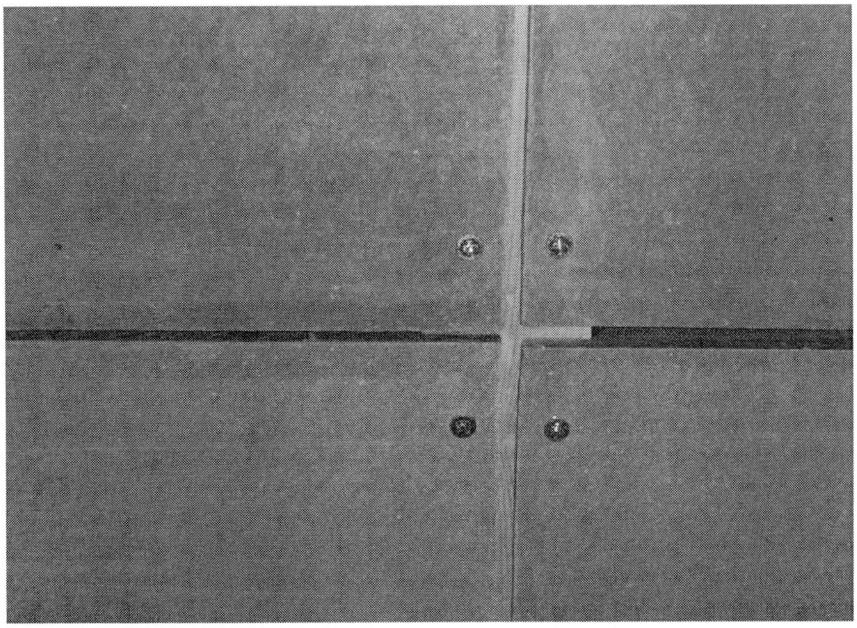

Fig. 3. Detail of gaps between CEMVIN boards of external ventilated cladding.

Fig. 4. View of the external cladding, gaps are sealed and covered by aluminium tape.

6 Results of measurements

Graph 1. The first group of measurements – the influence of the external ventilated cladding on sound insulation properties of structure with Steico Flex thermal insulation.

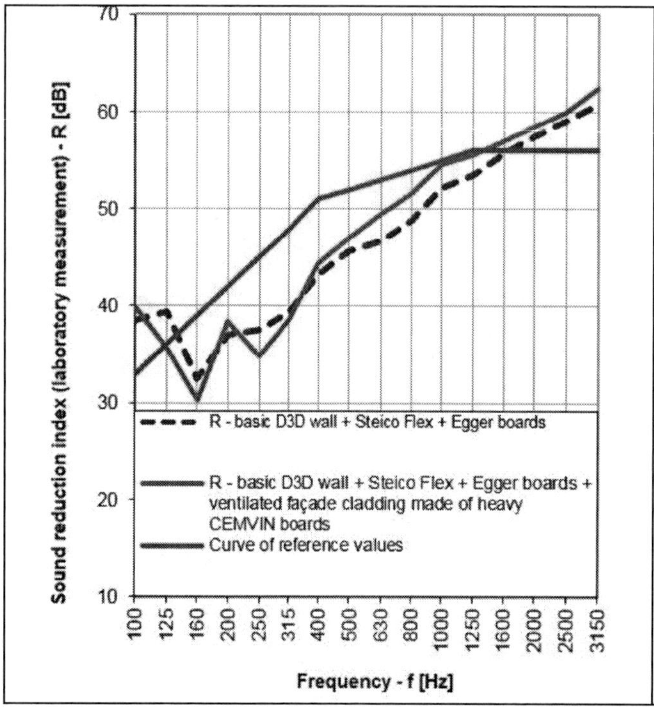

The single-number value of weighted sound reduction index and relevant spectrum adaptation terms laboratory measured for basic D3D wall insulated by Steico Flex and covered with Egger boards:

$$R_w(C;C_{tr})=49(-1;-5)$$

After application of external ventilated cladding made of heavy CEMVIN façade boards, the value changed followingly:

$$R_w(C;C_{tr})=49(-2;-5)$$

Graph 2. The second group of measurements – the influence of the external ventilated cladding on sound insulation properties of structure with ISOVER UNI thermal insulation.

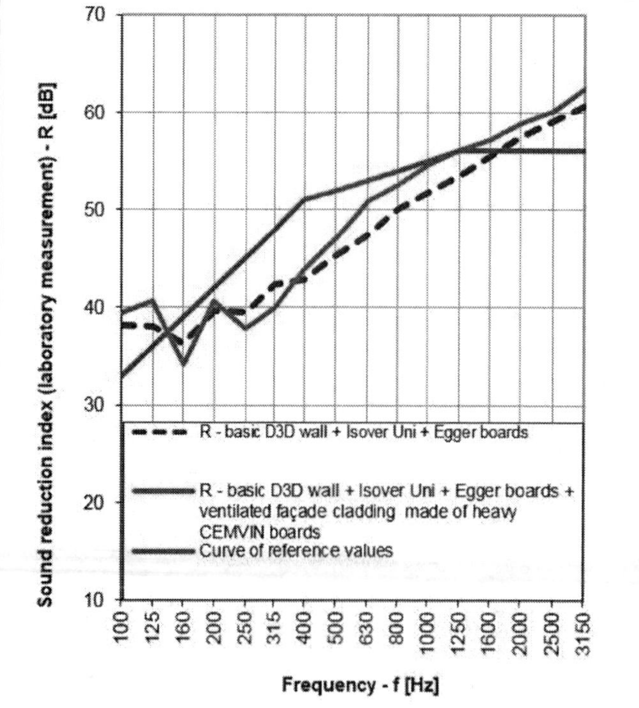

The single-number value of weighted sound reduction index and relevant spectrum adaptation terms laboratory measured for basic D3D wall insulated by ISOVER UNI and covered with Egger boards:

$$R_w(C;C_{tr})=50(-1;-4)$$

After application of external ventilated cladding made of heavy CEMVIN façade boards, the value changed followingly:

$$R_w(C;C_{tr})=51(-2;-5)$$

Graph 3. The third group of measurements – the influence of sealed external cladding gaps on sound insulation properties of structure with ISOVER UNI thermal insulation.

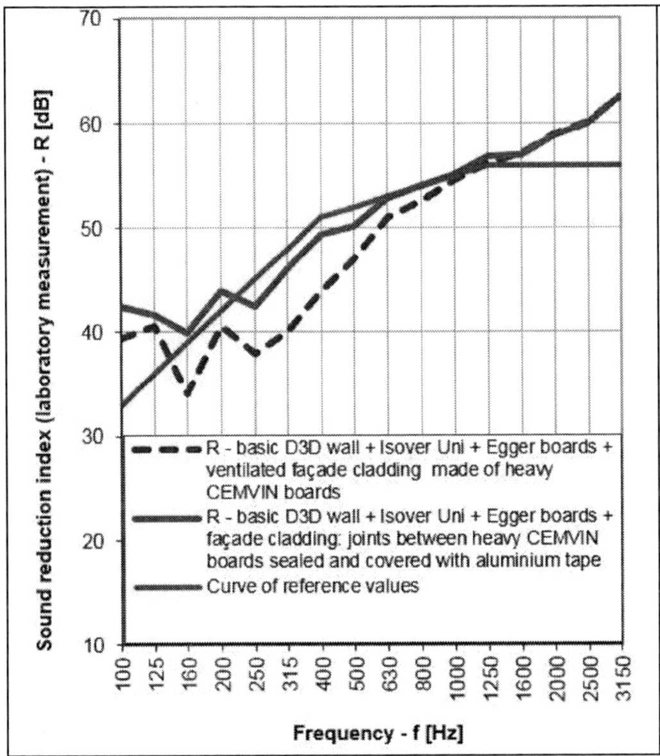

The single-number value of weighted sound reduction index and relevant spectrum adaptation terms laboratory measured for basic D3D wall insulated by ISOVER UNI and covered with Egger boards and heavy CEMVIN boards ventilated facade:

$$R_w(C;C_{tr})=51(-2;-5)$$

After sealing and covering gaps between heavy CEMVIN façade boards with aluminium tape, the value changed followingly:

$$R_w(C;C_{tr})=54(-1;-4)$$

7 Conclusions

Generally, the massive glued laminated timber construction insulated with Steico Flex showed the single-number value of weighted sound reduction index by 1 dB lower than the construction insulated with ISOVER UNI. Both constructions showed a significant improvement of laboratory sound insulation at all measured frequencies, compared to uninsulated constructions. This phenomenon has been previously described by the authors for example in [5].

It is important to mention that for structures described in this article, precise insulation execution has a fundamental influence on sound insulation properties. No major leaks or errors are allowed in the sample envelope wall.

Focused on the influence of external ventilated façade with free gaps, i.e. free joints between façade boards, the benefit of such construction is approx. 0 – 1 dB of the total laboratory sound insulation value, while the uncertainty is approx. 2 dB. Therefore, ventilated façade made of heavy façade boards has no unequivocal influence on the R_W value. The influence on the curve of sound reduction index R is apparent in the graphs. Nevertheless, the opinion that ventilated façade improves the sound insulation is still widespread among designers and builders. Consequently, faulty design leads to envelope wall defects.

After sealing and covering the gaps between façade boards with aluminium tape, the R_W value increased by 3 dB. Thus, if the gaps were covered, for example by different board installation method, the resulting construction could exhibit better sound insulation properties. Unfortunately, the ventilated façade needs vent inlets and outlets for proper function that act as acoustic weak point.

For a completely correct assessment of individual influences on sound insulation properties, more sample walls and an extensive data set would be necessary.

This article was created thanks to cooperation of the Research Institute for Building Materials and the Faculties of Civil and Mechanical Engineering of the Brno University of Technology, concretely internal project 134624-RO-10/2013 and Project of University specific research FAST-S-16-3345 "Verification of behaviour and development of structures and structural elements and combined structures from wood and reinforced concrete, Faculty of Civil Engineering, Brno University of Technology.

Let us thank especially to Doc. RNDr. Marie Vaňková, CSc., RNDr. Anna Ryndová, Ph.D. and Doc RNDr. Miroslav Doložílek, CSc., for assistance with realization of the measurement - complete information on measurement are included in [6].

References

1. *ČSN 73 0532, Acoustics – Protection against noise in buildings and evaluation of acoustic properties of building elements – Requirements* (Czech Office for Standards, Metrology and Testing, 2010)
2. *ČSN EN ISO 140-5. Acoustics - Measurement of sound insulation in buildings and of building elements - Part 5: Field measurement of airborne sound insulation of façade elements and façades* (Czech Office for Standards, Metrology and Testing, 2010)
3. Z. Fišarová, L. Kalousek, Akustika **22**, 12-19 (2014)
4. P. Valenta, Utility model no. 15073 U1, ČESKÉ DŘEVAŘSKÉ ZÁVODY PRAHA, A. S., PRAHA, CZ, Obvodový plášť zateplené budovy (External cladding of thermal-insulated building), available on http://www.upv.cz/cs/sluzby-uradu/databaze-on-line/databaze-patentu-a-uzitnych-vzoru.html
5. Z. Fišarová, L. Kalousek, M. Frank, Akustika **24**, 16-24 (2015)
6. *Protocol no. 35-2013 titled "Measurement of Laboratory Airborne Sound Insulation of external walls", issued by Acoustic Laboratory of the Institute of Physical Engineering, accredited as ČIA, No. 1016.* (2013)

Defects, faults and accidents of contemporary constructions

Michal Kraus[1,*], *Terezie* Vondráčková[1], and *Vladimír* Nývlt[1]

[1] Institute of Technology and Business in České Budějovice, Department of Civil Engineering, 370 01 České Budějovice, Czech Republic

Abstract. The paper deals with the defects, faults and accidents of contemporary constructions. The increasing number of constructions provides higher probability of occurrence of the defects. The defect is a lack of structure, which is caused by faulty design, inadequate or imperfect execution. The defect may affect the performance of the structure. The fault is the temporary or permanent loss of ability to perform its function. A defect may be causing disorder of structure. A special attention is paid to the defects and disruptions of buildings, which caused death or serious injuries (human or animal), safety hazards or major property damage. The aim of the paper is a description and analysis of data from the Construction and Technical Prevention System (CTPS) in the Czech Republic in several last years. It is necessary to take measures to prevent serious or excessively repetitive defects of buildings.

1 Introduction

Building production fluctuates in the Czech Republic over the last decade. Performance of construction industry increases up to 2008, followed by a fall. Re-starting the construction production is evident since 2014. The construction sector is constantly expanding. The more buildings, the higher the probability of occurrence of a defect or fault is. The increasing number of buildings is not the only factor increasing probability of occurrence of defects. Building defects are not only a problem of old buildings. Also, new buildings can have defects [1]. The continuous monitoring and analysing of repeated and serious incidents is in the public interest for prosperity and development of society. Other important factors are quality of the realization, quality of used building materials, technical equipment, proper use and regular maintenance. Every human activity is encumbered with some risk and the unavoidable error rate. This risk represents risk of injury, damage, malfunction or accident. The buildings sector is no exception [2]. The risk analysis is a universal method for assessment of risk of defects, failures and collapses [3-5].

Defect can be defined as a lack of construction due to incorrect design, inadequate or imperfect execution decreasing performance of structures. The defect is not a change of construction compared to the original state. The defect of building can may have technical and factual consequences. The fault is temporary or permanent depletion capabilities of

* Corresponding author: info@krausmichal.cz

construction. The fault is a change of construction compared to the original state. Faults may occur as a cause of the defect. Fault of construction has technical consequences. Accidents defines structural damage requiring repairs.

The Figure 1 shows the basic classification of the defects. In terms of function of construction, defects can be divided into serious and insignificant. The risk of injury, death or the emergence of substantial financial damage is characteristic of serious defects. According to the visibility, there are apparent defects and hidden defects. Apparent defect can be detected during routine maintenance or during the visual inspection. Hidden defect is not detectable by routine inspection or by maintenance [2].

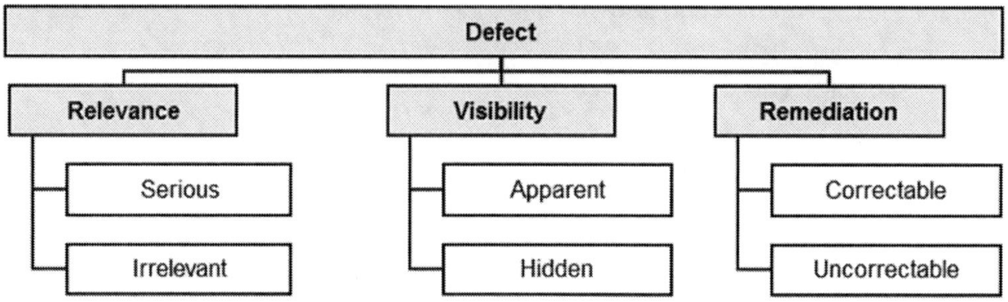

Fig. 1. Basic defect Classification [6].

2 Source data and its structure

The Construction and Technical Prevention System (CTPS) monitors serious or repeated defects in construction. The Institute for Spatial Development is responsible for the operation and administration of the system under the supervision of the Ministry of Regional Development of the Czech Republic. The CTPS is open access database. The system has been in operation since 2006. The Construction and Technical Prevention System monitors incidents that can be characterized by death of a person or animal, threat to life and health of people or animals, threat to the safety of the building and its equipment and by the emergence of property damage (approximately 18,000 Euros and more). According to the Building Act No. 183/2006 Sb., construction manager, construction supervision, developers and owners of buildings are obliged to notify the aforementioned incidents [6]. Other defects are not monitored by the system. The protocols have exactly defined form according to the Building Act. The structure of the analysed data is composed of year of occurrence, region, incidents, building category, functional unit/equipment and severity for each CTPS protocol.

Input data consist of a total of 516 reports recorded in the Construction and Technical Prevention System from 2010 to 2013. The sample includes the characteristic mix of contemporary buildings: residential, commercial, agricultural, industrial, educational, religious, government, transport building and infrastructure, site facilities and others from all regions of the Czech Republic. Overall, incidents are divided into nine categories of buildings.

In the context of the examined data sample, the statistical character "incident" represents situation that lead to defects, injuries or accidents in the construction practice. Generally, it is defined 6 kinds of incidents:

- Carbon monoxide poisoning, due to gas heaters, boilers and karma. In some cases, intoxication ends with the death of a person.
- Explosion, which was due to poor indoor installation or as yet unidentified causes. In most cases, it is the explosion of a gas pipeline as indoor installation of buildings.
- Fall from height, caving or injured animal.

- Fall from height, caving or personal injury.
- Fire, usually caused by careless use of open fire, the failure of indoor installation of building or by lightning. This category does not include the fire, which was created after the explosion.
- Structural collapse, building defect without explosion or fire leading to collapse or damage to construction.

Statistical sign "functional unit/equipment" describes the part of the structure characterized by a primary defect. The severity of the incidents is divided into five degrees from large financial damage without injury to the person's death (defined below).

3 Results

The sample is subjected to basic exploratory analysis. Absolute frequencies and relative frequencies are monitored for each group incidents. Incidents are analysed according to the territory, the building categories, and the functional unit of the building and according to the degree of severity.

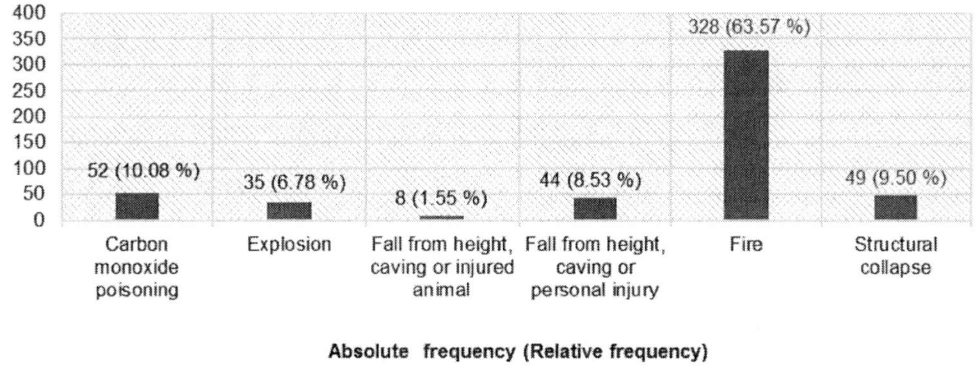

Fig. 2. Total overview of incidents.

The most common incident (modus) is a fire. It covers more than 64% of the sample. The main causes of fires are negligent conduct of building users as well as technical faults of functional units of constructions and building equipment. Relatively significant proportion also has carbon dioxide poisoning (10.08%), personal injuries (8.53%) and collapsed of structures (9.50%). Fall from height and fall into the ditch is the most common cause of injury. Vice versa, injuries to the animals has the smallest representation in data sample.

3.1 Territorial representations of incidents

The sample includes incidents from all 14 regions of the Czech Republic. As it is shown in Figure 3, the territorial representation of the incidents gives the total distribution of defects, faults and accidents in the Czech Republic. The most common area of monitored incidents is evident in region of Moravia-Silesia. Almost 16% of monitored incidents occurred in the Moravia-Silesia Region. The lowest density of incidents is region of Liberec (only 3.30%). It is necessary to look at the statistics territorial representation of incidents with considerable reserve due to a difference in the density of population and development in various regions.

 DOI: 10.1051/matecconf/20179303004

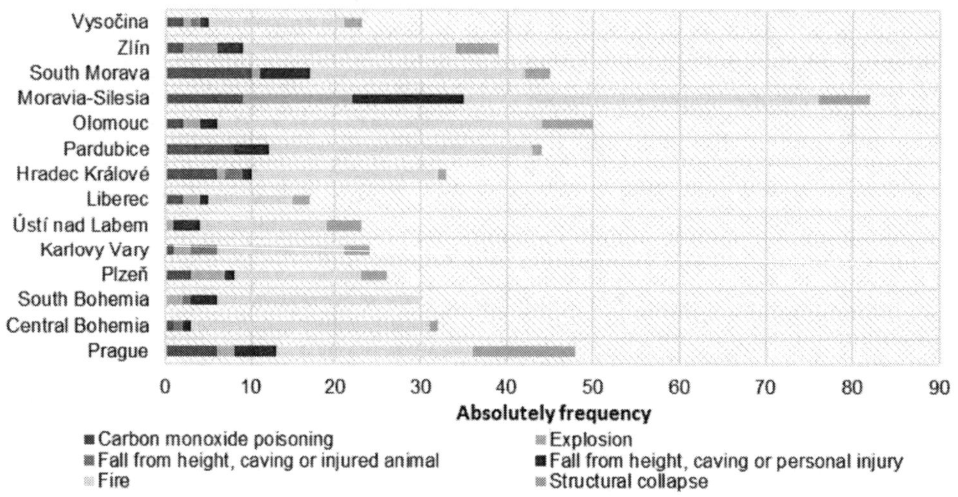

Fig. 3. Overview of incidents by regions in the Czech Republic.

3.2 Building categories of incidents

Building defects and incidents are common problems in any built structures. Altogether is defined nine categories of buildings (Figure 4). It is clear, the number of fires is the greatest in all categories of buildings. In the context of the occurrence of fires, residential buildings are most at risk. More than 55% of all fires are monitored in residential buildings. The buildings and storage are the second highest risk category. Conversely, the smallest number of incidents is monitored for educational, religious and government buildings. This differentiation confirms the diversity of fire protection requirements of various categories of buildings.

Site facilities are statistically significant for the high proportion of endangering the health and lives of people. The prevention and strict adherence to principles in the field of safety and health at work is an integral prerequisite for reducing the number of serious incidents in the future.

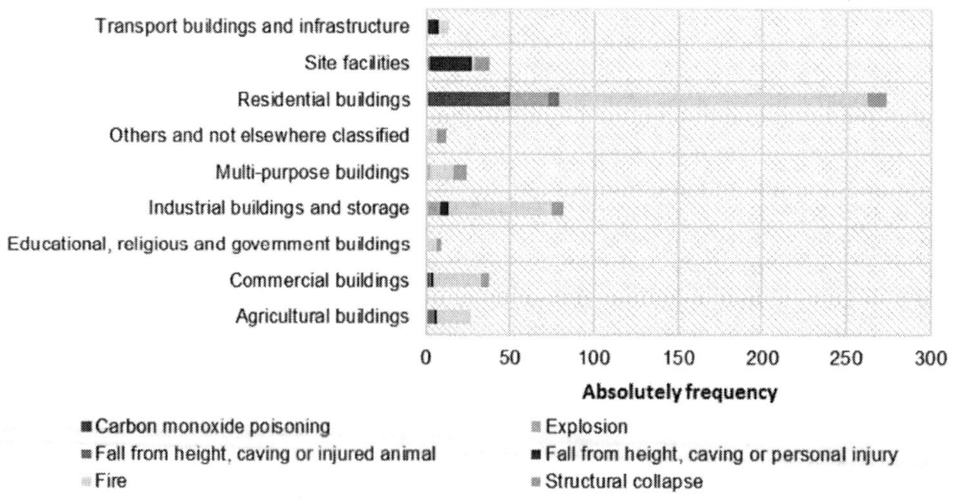

Fig. 4. Overview of incidents by building categories.

3.3 Functional unit and equipment of incidents

It is determined a total of 15 kinds of functional units or equipment (A-O) based on input data. Definition of functional units and their occurrence (relative frequencies) is described in Figure 4. Almost half of the data file (45.16%) consists of incidents caused by a combination of functional units and internal equipment as buildings. A significant proportion of incidents have also the roof structures (16.86%) and the heating systems (12.40%). Other functional units, equipment and indoor installations are equally risky for occurrence of defects and the threat to lives and health.

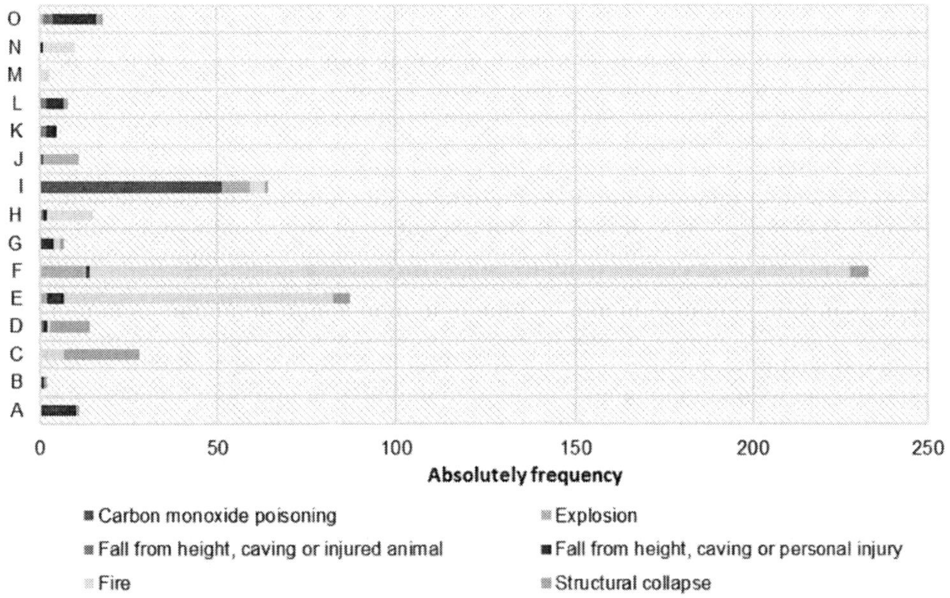

Legend:

Symbol	Functional unit/Equipment	Relatively frequency
A	Excavation work	2.13%
B	Foundation structures	0.39%
C	Walls and partitions	5.43%
D	Ceilings and horizontal structures	2.71%
E	Attic and roof	16.86%
F	Combination of constructions and interior equipment	45.16%
G	Transportation equipment	1.36%
H	Technological equipment and interior equipment	2.91%
I	Heating (indoor installation of buildings)	12.40%
J	Pipeline (indoor installation of buildings)	2.13%
K	Sewerage (indoor installation of buildings)	0.97%
L	Water supply (indoor installation of buildings)	1.55%
M	Ventilation (indoor installation of buildings)	0.58%
N	Electrical installation (indoor installation of buildings)	1.94%
O	Others and not elsewhere classified	3.49%

Fig. 5. Overview of incidents by functional units/equipment.

3.4 Severity of incidents

The input data include the severity of the incident. It is stated 5 degrees of severity from minor to serious. The first degree of severity (1) includes only significant financial loss without endangering of humans and animals. As mentioned above, the financial damage must be more than 18 000 Euros. Incidents with less financial damage are not analysed. The second degree of severity involves incidents threatening the lives and health of animals. The third degree of severity is defined by the deaths of animals. The fourth degree of severity involves the incidents of threats to human life and its health. The fifth degree (5) of severity, the most serious, represents the situations ending with death of one or more persons.

Figure 5 shows the final matrix of incidents and degrees of severity. Nearly 45% of all serious incidents end with serious financial damage. The rest defines cases of injury or death to persons or animals. More than 35% of incident represents injured people and 16 % of incidents ends by death. In terms of fatal accidents, the fire (43%) and fall from height or caving (21%) are the major cause of death. In the context of the categories of buildings, the residential buildings predominate with a share of 55%.

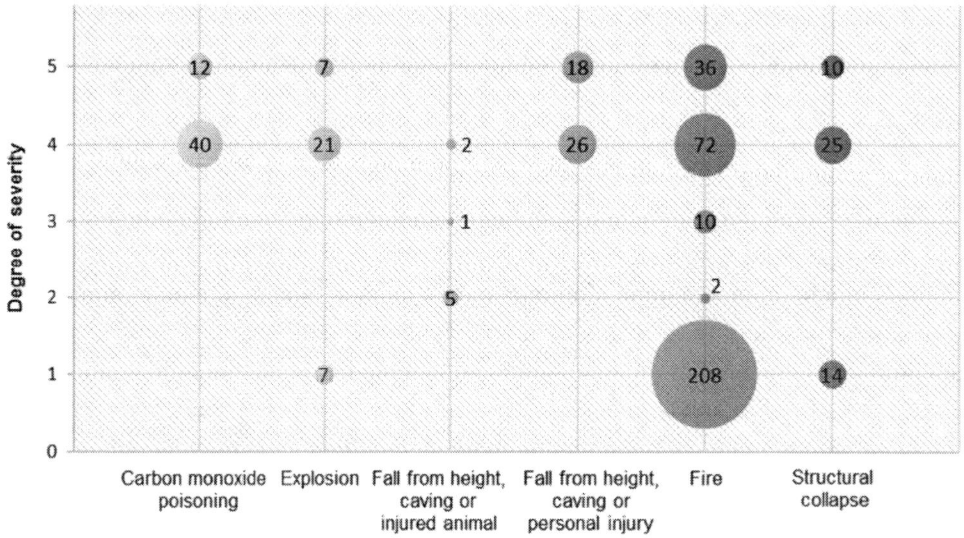

Fig. 6. Matrix of incidents and degrees of degrees of severity.

4 Conclusions

The aim of the paper is to evaluate and exploratory analysis of serious and recurring incidents of contemporary constructions. The analysis of incidents of contemporary construction brings new knowledge. More than half of the incidents the fire. It is necessary to pay attention to fire precautions and prevention not only during the construction phase and use, but also in the design phase. Negligence is a major cause of fires. The negligent conduct includes smoking, incorrect servicing of heaters and neglect of safety regulations. It is necessary to observe the principle of working with open flame, regular maintenance of technical equipment of buildings in order to minimize fires.

References

1. N.N.O. Bakrim M.A.O. Mydin, European Journal of Technology and Design **3** (2014)
2. K. Kubečka, *Transactions of the VŠB - Technical University of Ostrava, Civil Engineering Series*, 1, 27 (2010)
3. K. Kubečka, P. Vlček, D. Kubečková, AMR **899**, 535-538 (2014)
4. K. Kubečka, P. Vlček, D. Kubečková, D. Pieszka, AMR **1020**, 751-755 (2014)
5. M. Kraus, P. Bednářová, K. Kubečka AMM, **824** 657-665 (2016)
6. *Act No. 183/2006 Coll., on town planning and the building code (the Building Act)* (2006)

MATEC Web of Conferences **93**, 03005 (2017) DOI: 10.1051/matecconf/20179303005
Building Defects 2016

The human factor as a cause of failures in building structures

Terezie Vondráčková[1,*], *Věra* Voštová[2], and *Vladimír* Nývlt[1]

[1]The Institute of Technology and Business in České Budějovice, Okružní 10, 370 01 České Budějovice, Czech Republic
[2] Czech Technical University in Prague, Faculty of Transportation Sciences, Horská 3, 128 03 Prague 2, Czech Republic

Abstract. Human error can be defined as any unintentional or inadequate decision, taken at any level in the hierarchy of an organization, which is, or was inappropriate in a given situation. Human errors can occur in all human activities across an organization - at a managerial, conceptual or technical level – in connection with mistakes in the construction project itself, mistakes made by the investor, mistakes made by those using the building, mistakes in supplier relations, mistakes in the maintenance of the structure, and others. The factors that can affect the reliability of those people making the decisions that lead to the mistakes include the quality of their education, their experience with stress, or lack thereof, as well as issues such as workload, fatigue, workplace ergonomics, working hours, social climate and private matters.

1 Introduction

All construction works must be carried out in compliance with the requirements of operational safety, economic efficiency, the environment and in certain cases national security. Safety is one of the basic requirements in modern society. Increasing demand for ever greater efficiency and larger volumes of construction work has seen the gradual introduction of new building technologies and construction techniques. In parallel with this development, increasing standards have been introduced with regards to the quality and safety of operations, in particular in terms of management and logistics. This security depends on the proper management of people and the operation of equipment, as well as the relationship that exists between them.

The reasons for emergencies and accidents on construction sites are well known and documented. These reasons negatively affect the formation and progress of work processes. It is therefore necessary to analyse the prevailing risk factors, or combinations thereof, for the future. At present, several models exist that try to identify and describe the critical factors and interactions between humans and their surroundings [1].

* Corresponding author: vondrackova@mail.vstecb.cz

DOI: 10.1051/matecconf/20179303005

2 Models for human factors

The most widely known model is the SHELL model, which was developed in 1972 by Prof. Edward, and subsequently modified in 1975 by Prof. Hawkins. The word SHELL is derived from the initial letters of the model's components (see Figure 1). The components are: **L** (Liveware) - the people who use technical equipment, regulations, rules, etc. in their work and who are as individuals at the centre of interest; **H** (Hardware) – the equipment and functional systems; **S** (Software) - the rules, procedures and written documentation; **E** (Environment) - the situation in which the L-H-S system must function – the people with whom the individual at the centre of interest is in a relationship with (Liveware) and the working environment (Hardware and Software) in which the interactions take place. The human is an essential and critical element of the management of individual processes.

Fig. 1. SHELL model [2].

Another general model that characterizes the role of mistakes at several levels of an organization is the Reason model, alternatively called the Swiss cheese model. Under the model, human mistakes are considered to be a managed risk. If a mistake occurs at a certain level within an organization, and it is not detected and eliminated at another level, this may lead to a critical situation. One of the most sophisticated ways of working with data from road traffic accidents, engineering and manufacturing processes in the Czech Republic is mathematical modelling based on the Bayesian approach. Bayesian statistics is a branch of modern statistics, which works with conditional probabilities and allows for the more precise formulation of hypotheses because other relevant factors are taken into account. It is used very extensively where people work with uncertain knowledge [3].

3 Risks and causes

3.1 Risk sectors in the construction industry

In a report compiled by the State Labour Inspection it states that the most work-related accidents (43.1%) and fatalities occur in the manufacturing sector. High-risk industries include construction, transport and storage. Curiously, in 2013, the most common source of fatal work injuries involved ground vehicles and other means of transport, followed by buildings, structures and surfaces. In total, there were 25 fatal work injuries.

3.2 Causes of work accidents

In the majority of cases, work related accidents occur due to poorly or inadequately estimated risks. About 40% of all fatal accidents are caused by improper organization and safety at work, a lack of awareness among employers, dangerous work practices, actions undertaken without permission or despite an explicit prohibition, or by staying in a hazardous environment. These are reasons for making employers put greater emphasis on staff training, in particular in the area of work safety.

The reliability of any structure is defined as its ability to meet the requirements set out under certain conditions throughout its lifecycle. The safety of any building structure is not only affected by the natural properties of new materials, climate conditions and the like, but also to a large extent by serious mistakes during the development and construction process (see Table 1). Verifying the reliability of a structure on the basis of permitted load or through various methods, be they empirical, probabilistic, or on the basis of partial factors etc., significantly increases the complexity of calculations and therefore the economic burden [4].

Table 1. Causes of failure.

Draft	Implementation	Operation	Other	Human Error	Load
20%	50%	15%	15%	80%	20%

4 Work safety and the human factor

In today's modern and often hectic times, there is huge pressure from all sides not only on workers, but on mankind as a whole. This pressure creates many opportunities, but is also responsible for both psychological and physical problems, the latter being most closely associated with it.

Occupational safety and health has become an important task for every developed society and has become an essential part of working life and the environment. As we continue to create an effective system of prevention both in the technical and occupational hygiene senses, whose purpose is to protect human health and life, it is necessary for every worker to acquire the knowledge and moral sense and responsibility to implement it. In the construction sector, which is very specific in terms of its organization, operations and performance of work, the requirements for the safe performance of work are so extensive that they require special attention. After all, the underestimation of a risk, or the irresponsible behaviour of employees, can result in work accidents that may in some cases be so severe that they can cause permanent injury or even death.

4.1 Analysis of OSH in specific construction companies [5]

For the purposes of this analysis two construction companies were approached. They agreed to cooperate on the study on the basis of anonymity. Their names have therefore been changed to BUILDFULL and STAVOBRAVIA respectively. Both companies have branches throughout the Czech Republic and are considered to be large companies with a primary focus on transport and construction. Both companies concentrate on environmental protection and sustainable development.

BUILDFULL was established almost 150 years ago, employs approximately 1,100 workers, and has annual profits of approximately CZK 5 billion. The company operates in the fields of transport and civil engineering.

STAVOBRAVIA was established almost 80 years ago, employs approximately 1,000 workers, and has annual profits of approximately CZK 4 billion. The company operates in the fields of transport and civil engineering.

Both companies put a high emphasis on quality, environmental protection and reliability. For research purposes the focus was on issues of occupational safety and health (OSH) and on accident rates.

4.1.1 Work methodology

The work methodology was divided into two parts. The first part consisted of an anonymous questionnaire. The questionnaire was distributed to the employees of the companies who agreed to participate in the study. They were given sufficient time to complete it. The purpose of the questionnaire and the reason for processing it was given in the introduction. The first few questions referred to general information on the respondent, such as age, gender, etc. The questions that followed were structured in such a way that it was clear that the issue at hand was occupational health and safety being determined between the respondent and their employer. The questions were clearly formulated so that they could be easily understood and the response unequivocal.

The second part was an analysis of documents that the responsible staff in the surveyed companies provided for the research.

The results of the anonymous questionnaire were placed in tables and processed for the analysis of the data. The data was subsequently converted into the form graphs for better presentation and evaluation. By comparing the graphs an assessment was made of how the two companies act in terms of OSH. The same principle was applied to the analysis of the documents, from which emerged a clear evaluation of the factors under scrutiny.

4.1.2 Anonymous survey

In total, 30 employees in each of the two companies were sent a questionnaire relating to OSH at their employer. In total, 28 questionnaires were returned from BUILDFULL, and 30 questionnaires from STAVOBRAVIA. The questionnaires contained 10 questions.

1. *Your gender.* In BUILDFULL the sample consisted of 64% men and 36% women; In STAVOBRAVIA the sample consisted of 50% men and 50% women.
2. *Your age.* The largest age group was between 36-55 years, with the lowest proportion of respondents in the age group 25-35 years. The results were almost identical for both companies.
3. *Interested in the issue of OSH.* If we want to effectively reduce accident rates, it is important that the issue of occupational health and safety is approached by the staff in a responsible and positive manner. The results are presented in Figure 2. The results show a lack of interest or less interest in OSH within the companies. This may be the result of apathy towards training, a lack of motivation on the part of the employer or the unwillingness of employees.
4. *The role of OSH.* The results are presented in Figure 3.
5. *Health and safety training at your employer?* For both surveyed companies, 68% regularly participate in relevant training.
6. *The impact of the training.* In BUILDFULL 70% of the respondents considered trainings beneficial. The employees of STAVOBRAVIA were all satisfied.
7. *The frequency of training.* The results show that 10% of the 58 respondents were dissatisfied. This is therefore clearly an issue that both companies need to address.

DOI: 10.1051/matecconf/20179303005

8. *First aid kit in your workplace.* Only one person from both companies knew whether there was a first aid kit in their workplace. This shows gross ignorance of one of the basic principles of safety, which must be tackled.
9. *Identifying and preventing OSH risks.* In both companies, 15% of employees ticked the answer "do not know". It is clear that both the employees and employers are at fault on this issue.
10. *Critical day of the week.* The critical days were Mondays and Fridays.

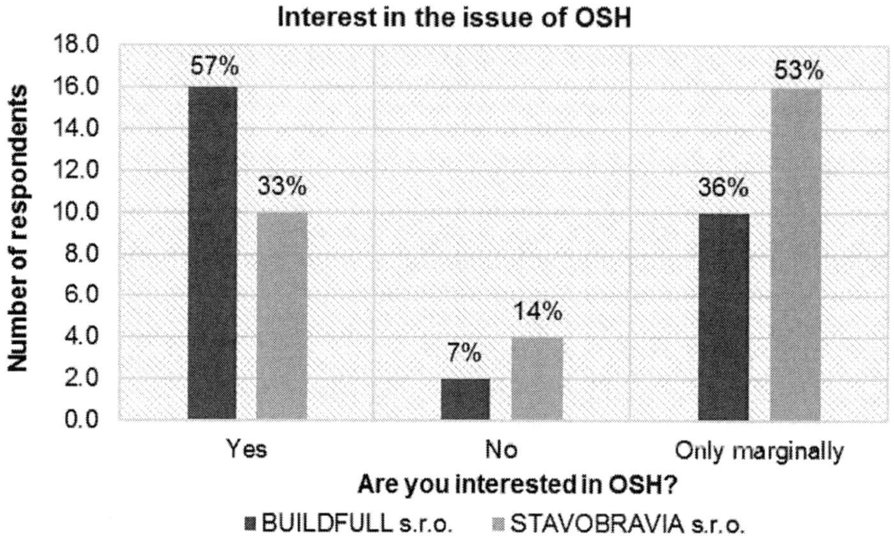

Fig. 2. Results of interest in the issue of OSH.

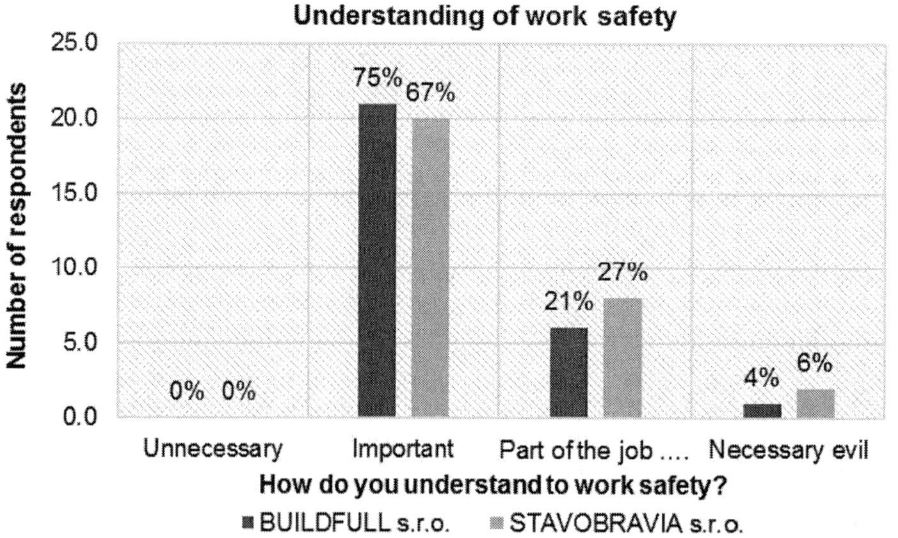

Fig. 3. Results of understanding of work safety.

4.2 Statistical analysis

The results in this section are based on an analysis of the occupational injuries over the past 6 years. Due to the similarity of the two companies in terms of the number of employees and their organizational structure, individual areas can be compared very well.

1. *The number of accidents in the period 2010-2015.* For both companies, consideration was given to the number of employees in each evaluated year, with the number of accidents and the frequency of accidents being given per 100 employees. The results are presented in Figure 4.

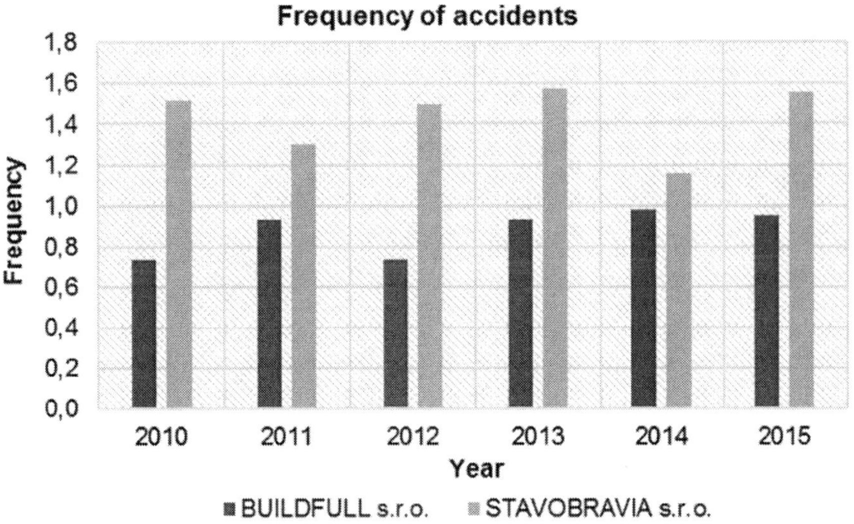

Fig. 4. Frequency of accidents from 2010 to 2015.

2. *The costs of purchasing personal protective equipment (PPE).* The results of the analysis show that the average annual cost for the purchase of PPE is approximately CZK 2,500 per employee per company.
3. *The number of missed working days per year in period 2010-2015.* In BUILDFULL the number was consistently around 350 days per year, whereas in STAVOBRAVIA it ranged from 300 to 400 days per year.
4. *The number of incurred injuries in the period 2010-2015 according to the days of the week.* From the research, it was clear that the most critical days are Mondays and Fridays. Saturdays and Sundays are the days with the minimum risk of accidents due to the fact that there are fewer workers on sites.
5. *The division of occupational injuries by age group.* In BUILDFULL the most accidents occurred in the age group 51 years and over. This result provides evidence of the fact that although the older should be more experienced and capable of recognizing and estimating risks in their activities, the opposite is true and that the situation is very often underestimated. The situation was similar in STAVOBRAVIA, where most accidents occurred in the age group 41-50 years. In both companies, the least accidents were recorded in the age group 18-30 years.

4.3 Discussion

The willingness on the part of the two companies to participate in the anonymous questionnaire survey shows that both companies are interested in the issue of OSH and deem it important and a part of their responsibilities. The results of the analyses show that both companies need to focus more on the concept and the content of the health and safety training they provide and, for example, use surveys to determine which information, notices or other aspects of the training the staff lack. Depending on these findings, the awareness of the employees should be raised. This should or could be part of regular OSH training sessions, or could possibly be the subject of one-off information sessions regarding the operation and facilities of the individual companies. The companies also need to thoroughly analyse their accident rates i.e. focus on the nature of individual accidents in order to reduce future risks and making it easier to prevent accidents. The research shows that although it might seem that older workers should be more experienced and more able to estimate the potential risks, the opposite is true. As a response to this, companies should concentrate on workers in the age group 41 years and over this group.

5 Conclusion

The human factor is a concept that is increasingly encountered both in the professional workplace as well as in the media. It is a "set of characteristics and the abilities of a human being, mainly psychological and physiological, that in some way in a given situation affects the performance, efficiency and reliability of the system of work." It is necessary to take into consideration that the concept of the human factor not only relates to individuals, but to groups, which determine the level of security (both in a company and in society) in terms of the depth of corporate culture, safety culture and societal culture. Managers in companies should therefore focus on this phenomenon because in it lies one of their weakest links.

Occupational safety and health is a very broad field that involves extensive regulations, standards, laws, etc. If it is in any way possible to eliminate risks and protect human health and life, it is necessary to develop this sector and strive for the greatest fulfilment of its objectives.

References

1. S. Szabo, Journal of Applied Economic Sciences **10**, 1153-1159 (2015)
2. Lidský činitel v letecké dopravě, available on http://projekt150.ha-vel.cz/node/117
3. R. Štecha, *Human Factors Classification in Air Traffic Control* (Disseration thesis, Czech Technical University in Prague, Faculty of Transportation Sciences (2013)
4. T. S. Ferry, *Modern Accident Investigation and Analysis* (New Jersey, John Wiley and Sons, 1988)
5. L. Homolková, *Analýza BOZP ve zvolené stavební firmě v časovém horizontu 5 let* (Bachelor thesis, Institute of Technology and Business in České Budějovice, 2016)

Vapour-tight layer and its effect on surface temperatures of curtain walling

Petr Kacálek [1,*] and *Tomáš* Petříček[1]

[1]Brno University of Technology, Faculty of Civil Engineering,
Institute of Building Structures, Veveří 95, 602 00 Brno, Czech Republic

Abstract. The primary function of a vapour-tight layer is to reduce or prevent the spread of humidity from the interior to other layers of the surrounding structures. This would otherwise result in condensation of (water) vapour due to a drop in temperature. In some cases, the behaviour of real structures differs from the structure modelled using a computational programme. This paper deals with the factors that can affect the features of a structure, specifically the internal and external surface temperature of curtain walling. The heat reflection effect of a vapour-tight layer was investigated on a model panel of curtain walling. An inadequate surface temperature can have a negative impact in that it promotes a biotic attack. This results in surface degradation which affects the continuous layers in the adjacent structures.

1 Introduction

The interest in prefabricated wooden houses has grown over time. They have recently become the most popular type of residential house. In this type of building, it is necessary to not only protect the timber elements, but also a number of specific constructional details, as well as provide a vapour-tight barrier from the external wall. Vapour-tight barriers in wooden sandwich constructions are usually one of various types of PE foil. Another option is to use PE foil with a thin layer of aluminium for effective heat reflection. The application of the appropriate vapour-tight layer can have a significant influence on the final surface temperature of the structures.

Due to the sealing of the building´s envelopes, surface humidity from condensation or even the growth of mould on the structure´s surface is often encountered. In the majority of cases the cause is the inappropriate use of interior materials. However, there are other, less significant factors that can also affect the behaviour of the structure.

An inadequate surface temperature can have a negative impact in that it promotes a biotic attack on the structures of a prefabricated wooden house. The result is surface degradation which affects the continuous layers in adjacent structures. In some cases, the

[*] Corresponding author: kacalek.p@fce.vutbr.cz

behaviour of the real structure can differ from the behaviour of the structure modelled using a computational programme.

The research conducted in this paper is based on a sandwich test panel (made up of two parts) constructed using real wooden house curtain walling with two types of vapour-tight foil (standard PE foil and reflective multilayer bubble foil) and a plasterboard lining. The surface temperature was measured using thermocouples placed in comparable positions on both halves of the panel i.e. with the standard or reflective vapour-tight foil. The effect of using an aluminium reflective layer on one of the parameters was monitored in order to evaluate the quality of the indoor environment – internal surface temperature of the curtain walling.

2 Methodology and course of the experiment

The testing of the properties of the vapour-tight layers involved contact measurement of the surface temperature of the two halves of the test panel using thermoelectric surface sensors (thermocouples). During the contact measurement, the temperature sensors were in direct contact with the surface being measured. The method for attaching the sensor to the surface was chosen so that the temperature of the surface was not influenced by the sensor and the sensor was in direct contact with the surface. The measurement process was split into two phases. In the first phase, the measurement was only taken on the plasterboard lining of the curtain walling. In the second phase, the subject of the measurement was the complete structure i.e. the curtain walling of the prefabricated wooden structure and the plasterboard lining (see Figure 1).

Fig. 1. Curtain walling test panel of the prefabricated wooden building structure.

On one half of the test panel, a vapour-tight layer made of multilayer bubble foil with an aluminium reflective layer (REF) was applied, and on the other half of the panel, a vapour-tight layer without a reflective layer (STANDARD) was applied. During the measurement process, the transient temperature of the external environment was simulated.

3 Experimental measurements

The test panel, complete curtain walling and plasterboard lining, with the dimensions 3 x 1.5 m (see Figures 2 and 3) was installed in the test laboratory at the Institute of Building Structures. The two halves of the panel were tested by simulating the transient temperature, whereby the external temperature (Te) in the range of (-15) ±1°C to (-18) ±1°C and the internal temperature (Ti) 21±1°C. The monitored values (panel surface temperature, internal and external environment temperatures) were recorded at 10-minute intervals.

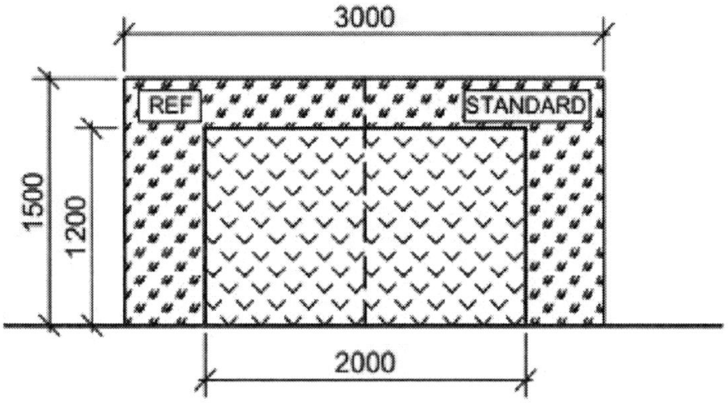

Fig. 2. (Two halves of the) test panel [1].

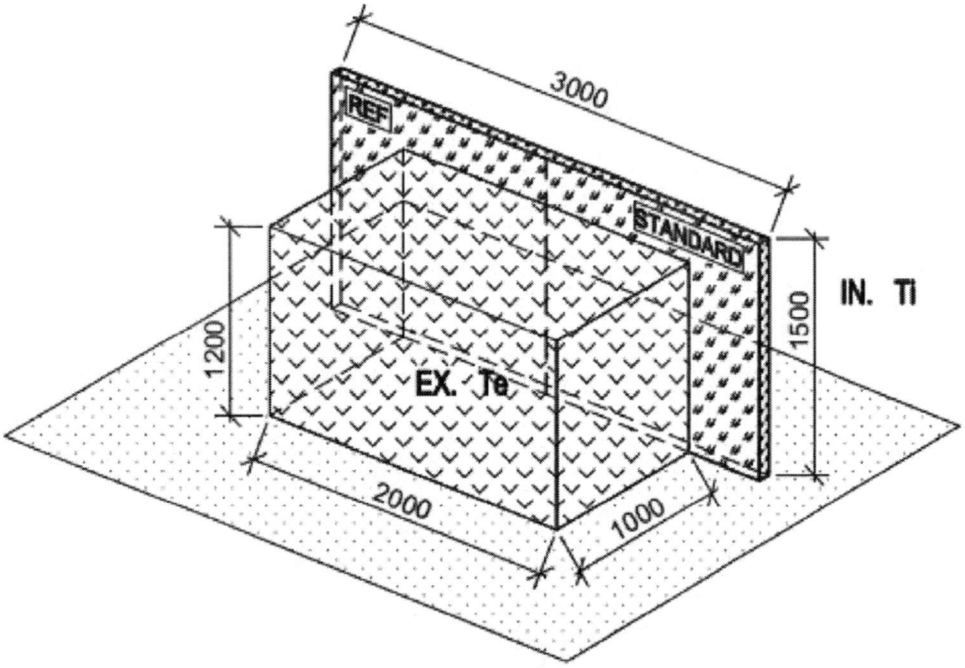

Fig. 3. 3D view of the test panel [1].

The structures of the two halves of the test panel are shown in Figures 4 and 5.

DOI: 10.1051/matecconf/20179303006

When applying the reflective vapour-tight barrier with aluminium layer, a non-ventilated air gap of at least 25 (30) mm thickness must be created in front of the barrier. This allows the reflection of the heat back into the interior.

For the two halves of the test panels, the structure of the interior plasterboard lining (see Figure 4), as well as the complete structure with curtain walling (see Figure 5), were constructed in the same way they are commonly installed in real structures for prefabricated wooden houses.

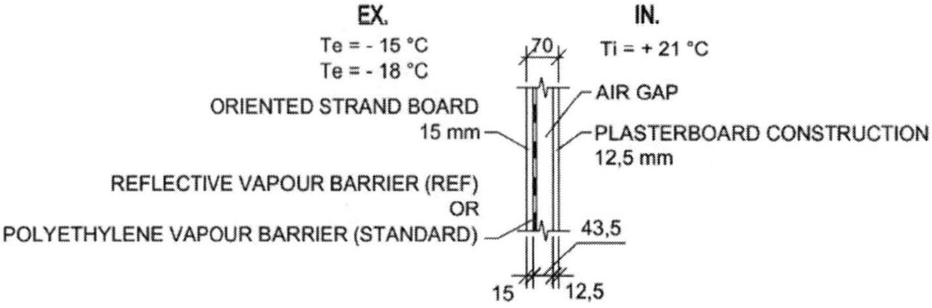

Fig. 4. Composition of test panel - plasterboard lining.

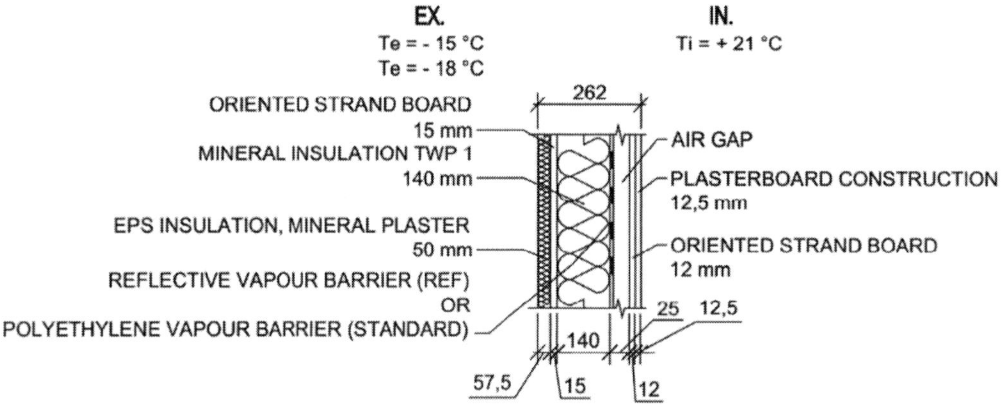

Fig. 5. Composition of test panel - curtain walling and plasterboard lining.

4 Experimental measurements results

The experiment provided four sets of results – for two types of barrier and two external temperatures. The results are presented in Tables 1 and 2.

The temperature difference observed on the external surface of the plasterboard lining with the reflective foil applied was about one third higher compared to the temperature measured on the wall on which only PE vapour-tight foil was applied. During the measurements, when the ambient temperature was Te = -15 0C (see Table 1), the temperature on the surface where the foil with the aluminium reflective layer was applied was 33.6% higher (from 21.7°C to 29°C). In other words, the application of the foil significantly increased the thermal resistance of the wall.

Table 1. Results of measuring surface temperature of plasterboard lining [2].

Cold side of the wall (EX.)			**Warm side of the wall (IN.)**		
Temperature Te (°C)	Surface temperature (°C)		Temperature Ti (°C)	Surface temperature (°C)	
	polyethylene barrier	Reflective barrier		polyethylene barrier	Reflective barrier
-15.0	-7.4	-13.3	21.0	14.3	15.7
-18.0	-9	-15.6	21.0	14.3	15.7

In terms of curtain walling test panel, the thermal insulator in the structure has a significant influence. The difference in temperatures measured on the surface of the curtain walling where the reflective foil was applied was negligible compared to the wall on which only the PE vapour-tight foil was applied. During the measurements, when the ambient temperature was Te = -15 ^{0}C (see Table 2), the application of the foil with the aluminium reflective layer increased the temperature difference from 32.42 ^{0}C to 34.02 ^{0}C i.e. by 5 %.

Table 2. Results of measuring surface temperatures of curtain walling test panel.

Cold side of the wall (EX.)			**Warm side of the wall (IN.)**		
Temperature Te (°C)	Surface temperature (°C)		Temperature Ti (°C)	Surface temperature (°C)	
	Polyethylene barrier	Reflective barrier		Polyethylene barrier	Reflective barrier
-15.0	-13.81	-15.33	21.0	18.61	18.69
-18.0	-15.86	-18.38	21.0	18.65	18.74

5 Conclusion

By experimenting with plasterboard lining and curtain walling test panels, the effect of a vapour-tight barrier with applied reflective aluminium layer was investigated. On the basis of the measured values, it is clear that the application of a multi-layer bubble foil with aluminium reflective layer does not significantly influence the internal surface temperatures of curtain walling with common heat insulation.

From the results of the measurements taken on the real curtain walling panel (see Table 2) it is clear that the application of the foil with the reflective layer has a significant effect on the external side of the wall where the temperature difference is ca. 1.5°C. The effect of the application of the reflective foil on the internal side of the wall is only slight compared to the total composition of the curtain walling, whereby the application of mineral heat insulation between the load-bearing columns has a significant effect on the thermal gradient.

The “light” vapour-tight layers do not significantly improve the thermal characteristics of the structure, especially the internal surface temperature which significantly determines the formation of biotic attack on the surface of the structure. Emphasis should therefore be placed on the incorporation of a high quality vapour-tight layer into the structure. The most important things are to avoid construction errors, to ensure perfect connections to passages, and other important details (e.g. threshold connector – [3]. etc.). Another important factor is that the vapour-tight layer must be of sufficient mechanical strength, which is influenced by the lining layer and the individual joints of the vapour-tight layers [4-6].

This paper has been worked out under the project No. LO1408 "AdMaS UP - Advanced Materials, Structures and Technologies", supported by Ministry of Education, Youth and Sports under the „National Sustainability Programme I”.

References

1. P. Kacálek, R. Smolka, T. Petříček, *Juniorstav 2011* (Akademické nakladatelství CERM, Brno, 2011)
2. H. Marková, P. Kacálek, J. Hejhálek, Českomoravský výběr (Vega s.r.o., Hradec Králové, 2009)
3. R. Smolka, J. Plachý, P. Kacálek, T. Petříček, AMM **824**, 188-195 (2016)
4. T. Petříček, P. Kacálek, R. Smolka, AMR **1041**, 71-74 (2014)
5. *CSN EN ISO 10211, Thermal bridges in building construction - Heat flows and surface temperatures - Detailed calculations* (Czech office for standards, metrology and testing, Prague, 2009)
6. *CSN EN ISO 13788, Hygrothermal performance of building components and building elements — Internal surface temperature to avoid critical surface humidity and interstitial condensation — Calculation methods* (Czech office for standards, metrology and testing, Prague, 2013)

Characteristics of the mechanically fastened joints of one-layer bitumen sheets

Tomáš Petříček[1,*], *Petr* Kacálek[1], and *Tomáš* Hlavačka[1]

[1]Brno University of Technology, Faculty of Civil Engineering,
Institute of Building Structures, Veveří 95, 602 00 Brno, Czech Republic

Abstract. The system of mechanic fastening is one of the most frequently used variants of roof membrane stabilization against the impacts of wind sucking. With the growing number of applications of mechanically fastened roofs the number of defects is also growing. We particularly see them in the joints of waterproofing membranes. The contribution is thus dedicated to examination of mutual dependence between the type of the carrier and the strength characteristics of mechanically fastened joints of these sheets. Bitumen sheets with different types of carrier used for one-layer mechanically fastened waterproofing membranes of flat roofs were chosen as samples. The above mentioned dependence is frequently stated in the practice, but is not explicitly documented. The aim of the contribution is to confirm or refute a hypothesis that a relation between the bitumen sheet carrier type and the strength of their joints exists. The results were arranged on the base of statistic evaluation of data obtained by laboratory measurement.

1 Introduction

The application of the single-layer mechanically fastened bitumen sheet system is a quick and progressive one, which requires high technological discipline. High quality bitumen sheets of at least 0.5 mm thickness are required as standard for the application of single-layer systems. The bitumen sheets also need to be of a sufficient quality that they are weldable and flexible over the long term. The sheets are mostly reinforced with a combination of PES fleece strengthened with glass fibres laid in one or both directions. Overlaps of at least 120 mm wide are recommended for the sheets.

Whole surface melting of the joint is essential to the waterproofing capabilities of the system as well as for the transfer of wind load forces. A critical point with regards to the wind load resistance of a joint is the presence of a fastening element.

2 Fastening of bitumen sheet joints

To ensure the joint is sealed correctly the surface area along the entire joint length of both sheets need to be heated to melting point. The fastening element is not placed in the centre

* Corresponding author: petricek.t@fce.vutbr.cz

of the joint but is located 10 mm (or 20 mm) from the edge of the lower sheet, whereby the entire surface area of the joint is able to take the strain of the fastening element. The effects of linear shearing (tear) force and the tangential forces (peel resistance) are therefore applied in varying degrees to the overlap over the entire surface area of the joint.

According to ČSN EN 13707 [1] manufacturers of bitumen sheets are obliged to produce them with sufficient resistance against tangential forces (Determination of peel resistance in joints according to ČSN EN 12316-1 [2]) and linear shear resistance when joined together (Determination of shear resistance according to ČSN EN 12317-1 [3], [4]).

A mechanically fastened bitumen sheet joint faces a combination of both the above stresses. As angle β grows (see Fig. 1), the level of peel stress grows, whereas where angle β (Fig. 2) is minimized, the linear shear force (tear) grows. Within this context, the peel resistance test of a joint is of particular importance when evaluating the quality of the bitumen substance used [5].

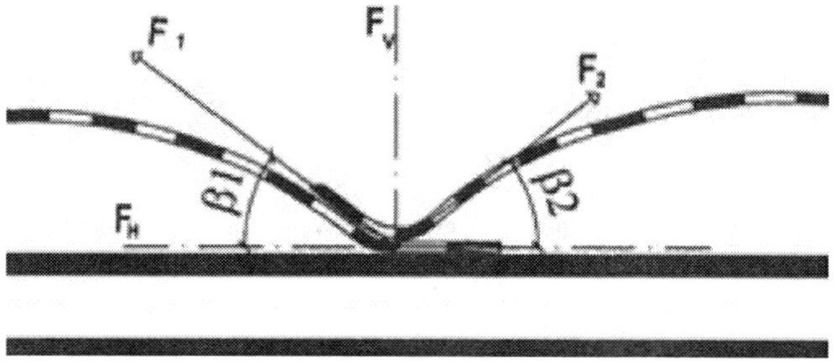

Fig. 1. Prevailing tangential (peel) force. Source: [6].

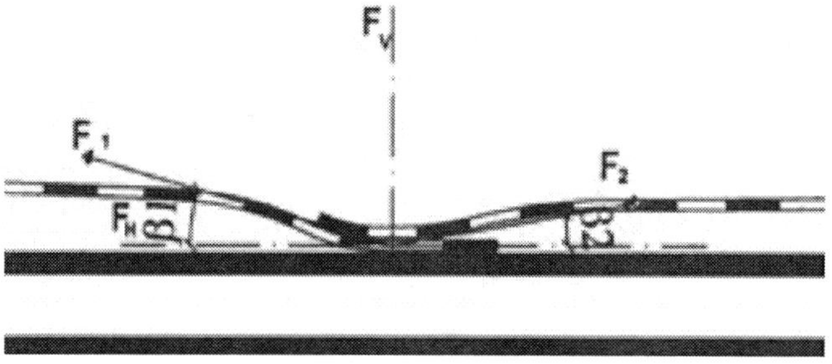

Fig. 2. Prevailing linear shear (tear) force. Source: [6].

The carrier in the bitumen sheet guarantees the strength properties, volume stability, it forms the base of the bitumen sheet during production and also traps and transfers tensile stresses and redistributes the stress caused by internal and external forces.

The resultant mechanical and strength characteristics of the carrier depend on the material base of the fibres, their mutual direction and on the proportion of the applied filler for their connection. Approximate comparison of the properties of carriers are shown in Table 1.

Table 1. Approximate comparison of strength and ductility of the most often used carriers Source: [6].

Carrier type	Square weight	Strength		Ductility	
	g/m^2	longitudinal direction [kN.m/$^{-1}$]	transversal direction [kN.m/$^{-1}$]	longitudinal direction [%]	transversal direction [%]
Glass fleece (Vlis)*	60-110	min. 3.6-8.6	min. 2.6-4.8	1.2-1.4	1.2-1.4
Glass woven fabric (rug)**	152-200	min. 18-21	min. 20-24	2-3	2-3
PES fleece***	120-250	8.6-19	4.6-14	20-35	23-37
Combined****	160-180	min. 9.6-13.3	min. 6.8-8.1	min. 23	min. 25

* Johns Manville, DH series ** Saint Globain Vertrotex
*** Johns Manville, 032 series **** Johns Manville, 054 series

Tests describing tensile properties provide comparison of the influence of the carrier on the overall strength parameters of a bitumen sheet: the highest tensile strength and ductility specified in the longitudinal and transversal direction. The graph in Fig. 3 shows comparison of mechanical properties of carriers and bitumen sheets with the respective carriers.

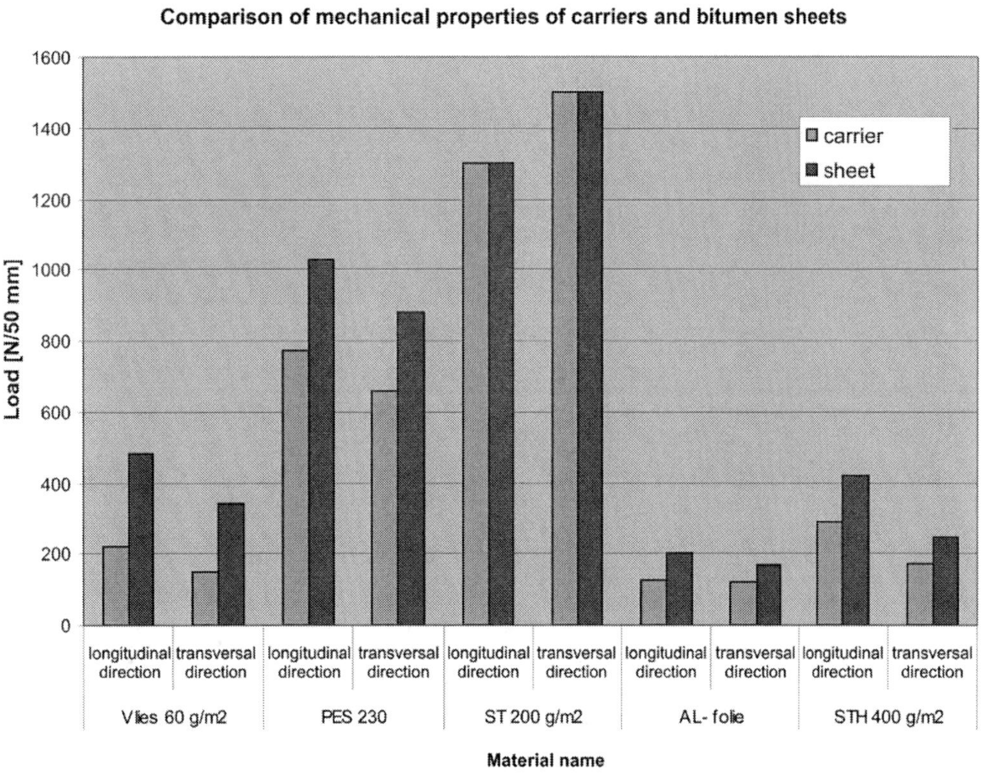

Fig. 3. Comparison of mechanical properties of carriers and bitumen sheets with respective carriers at the temperature of +20 °C Source: [6].

3 Experimental measurement

A pressure chamber was used in which the composition of a roof membrane from a real joint segment of mechanically fastened waterproofing could be subjected to defined stresses. This wind load test is defined in the Approval Guideline for systems of mechanically fastened waterproofing membranes ETAG 006 [7].

For the experimental measurements tests were carried out on a standard tensile testing machine. The specimens were roof membrane segments, not just the waterproofing membrane. Two metal fastening strips in the arms of the upper grip of the machine held the waterproofing membrane in place. These arms transferred the force of the applied stress to test material. A pad, to which the waterproofing membrane was fastened by thermal insulation, was clamped into the lower grip. As the lower grip of the tensile testing machine receded it caused stress on the joint and subsequently its rupture.

The test specimens were prepared according to composition variant S1 i.e. analogically for a roof without thermal insulation with the waterproofing membrane mechanically fastened to OSB board of 22 mm thickness. The fastening element to the hard base was a screw with a metal oval washer. The edge of the pressure washer of the fastening element was located 10 mm from the edge of the lower bitumen sheet. The use of the oval pressure washer was intentional because it provides better pressure on the lower sheet due to its larger surface area. As a result, the rupture of the test specimens was dependent on the parameters of the bitumen sheet rather than on the failure of the fastening element.

Three bitumen sheet variants were chosen as the waterproofing membranes:

1. Test specimen B - SBS modified bitumen sheet 5.2 mm thick, carrier: PES 180 g/m^2 reinforced longitudinally with glass fibres, flexibility at low temperatures: -25 °C, joint length 100 mm

2. Test specimen E - SBS modified bitumen sheet 5.2 mm thick, carrier: PES 230 g/m^2, unreinforced, flexibility at low temperatures: -25 °C, joint length 100 mm

3. Test specimen F - SBS modified bitumen sheet 5.2 mm thick, carrier: PES 180 g/m^2, reinforced on both sides with glass fibres, flexibility at low temperatures: -25 °C, joint length 100 mm

Fig. 4. Test specimen - composition variant S1. Source: [8].

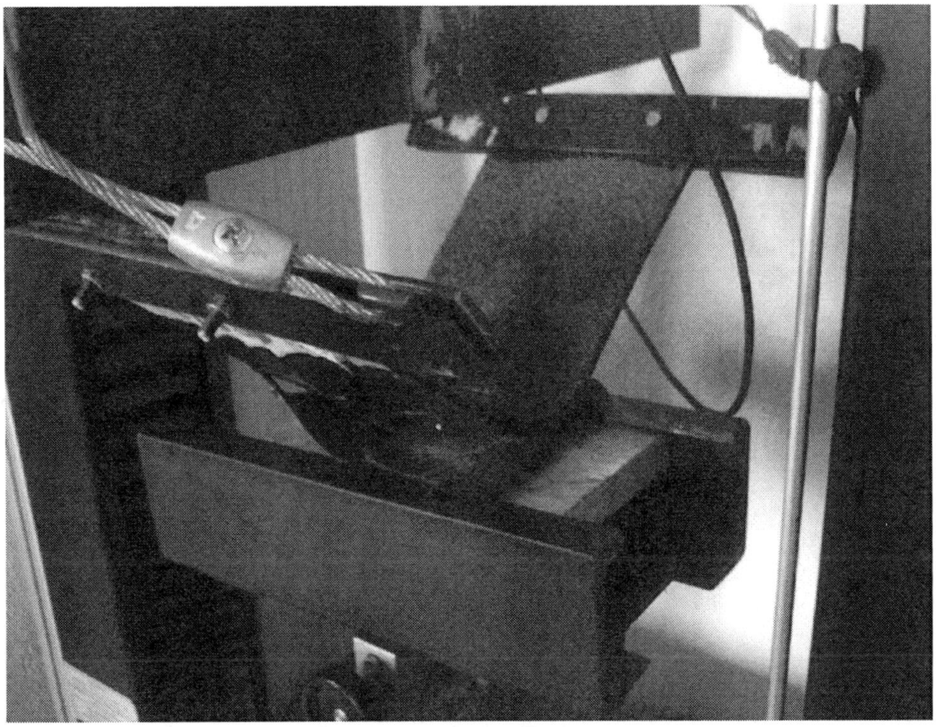

Fig. 5. Test specimen in the tensile testing machine during testing. Source: [8].

4 Evaluation

Approximately 10 specimens were used for each test sample. The final force is in the form $\bar{x} \pm \bar{\vartheta}(\bar{x})$, where $\bar{x}$ is the arithmetic average and $\bar{\vartheta}(\bar{x})$ is the probable measurement error. The latter was determined on the basis of the error theory for the deviations of the arithmetic average using the formula:

$$\bar{\vartheta}(\bar{x}) = \frac{2}{3}\sqrt{\frac{1}{n(n-1)}\sum_{i=1}^{n}(x_i - \bar{x})^2}. \tag{1}$$

5 Results of the experimental measurement and discussion

In the specimens with lower square weight (B and F) unsealing of the joint occurred and the test ended in tearing of the carrier on the base of the fastening element and in separation of the waterproofing layer from the base (Fig. 6). The strength of specimen E exceeded the range of the testing machine (max. tensile 2,000 N), the coherence of the specimen E was so high that no unsealing of the joint occurred in some of the samples (Fig. 7).

We can assume on the base of the results of the experimental measurement that the strength of a mechanically fastened joint grows with the square weight of the PES rug used in the bitumen sheet carrier. Specimen E (unreinforced PES rug, 230 g/m^2) achieved the highest strength values, the results of specimen B (PES rug longitudinally reinforced with glass fibre, 180 g/m^2) were by approximately 13 % lower. Specimen F (PES rugs reinforced with glass fibre on both sides 180 g/m^2) showed the lowest strength, the values were by approximately 20 % lower compared to specimen E.

Detailed results and comparison of the results of all the specimens are shown in Table 2.

Fig. 6. Typical damage of the joint on specimen B. Source: [8].

Fig. 7. No damage of the joint on specimen E is visible at the maximum tensile. Source: [8].

 DOI: 10.1051/matecconf/20179303007

Table 2. Comparison of the test specimens

Test specimen:	**B**	**E**	**F**
Bitumen substance:	SBS modified		
Flexibility at low temperatures:	-25 °C		
Sheet thickness:	5.2 mm		
Carrier:	PES longitudinally reinforced, 180 g/m2	PES 230 g/m2, not reinforced	PES reinforced on both sides, 180 g/m2
Overlap width:	100 mm		
The highest tensile strength:	longitudinally: 900 (± 200) N/50mm, transversally: 700 (± 150) N/50mm	longitudinally: 950 (± 200) N/50mm, transversally: 850 (± 150) N/50mm	longitudinally: 900 (± 200) N/50mm, transversally: 650 (± 150) N/50mm
Ductility:	longitudinally and transversally: 50 (± 10) %	longitudinally and transversally: 50 (± 10) %	longitudinally and transversally: 45 (± 10) %
Mean value:	**1,624.5 N**	**1,856.8 N**	**1,488.3 N**
Standard deviation:	127.6 N	97.4 N	131.6 N
Variation coefficient:	7.9%	5.2%	8.8%

Higher square weight of the PES rug increases resistance of a bitumen sheet against mechanical damage (tearing against the shank or the pressure washer of the fastening element) when exposed to stress. For combined carriers manufacturers present their total square weights, i.e. the PES rug including the reinforcement glass fibres. In the instance of the PES rug reinforced on both sides the proportion of the glass fibres in the total weight of the carrier is higher, so the weight of the PES rug itself is lower. This might explain why the bitumen sheet with longitudinally reinforced PES rug (specimen B) achieved higher strengths than the sheet with a PES rug reinforced in both directions (specimen F).

In the bitumen sheets reinforced with combined carrier no gradual cracking of the individual glass fibres that reinforce the carrier was visible during tensioning and before the rupture. In the case of the combined carriers the parameters depend on the particular type – square weight, location and connection of the reinforcement fibres to the PES rug affect the resulting strengths and ductility of the carrier.

The results of this comparison should however be understood in wider context – on the base of the above described results application of bitumen sheets with carriers of non reinforced PES rugs cannot be preferred. Application of combined carriers in single-layer mechanically fastened systems provides irreplaceable advantages.

6 Conclusion

The performed measurements showed that the strength of a joint of mechanically fastened bitumen sheet grows with the square weight of the PES rug used in the carrier of the bitumen sheet.

In combined carriers the total weight is lowered by the weight of the glass reinforcement fibres, which might lead to reduction of the final strength compared to an unreinforced carrier. Application of combined carriers in single-layer mechanically fastened systems however provides irreplaceable advantages.

This paper has been worked out under the project No. LO1408 "AdMaS UP - Advanced Materials, Structures and Technologies", supported by Ministry of Education, Youth and Sports under the „National Sustainability Programme I".

References

1. *ČSN EN 13707+A2, Waterproofing sheets and foils – Reinforced roof waterproofing bitumen sheets – Definitions and characteristics* (The Institute of Technical Standardization, Metrology and State Testing, Prague, 2009)
2. *ČSN EN 12316-1, Waterproofing sheets and foils – Part 1: Roof waterproofing bitumen sheets – Determination of peel resistance of joints* (Czech Standards Institution, Prague, 2000)
3. *ČSN EN 12317-1, Waterproofing sheets and foils – Part 1: Roof waterproofing bitumen sheets – Determination of shear resistance in joints* (Czech Standards Institution, Prague, 2000)
4. R. Smolka, T. Petříček, P. Kacálek, *enviBUILD 2014 - Buildings and Environment* (Brno, 2014)
5. T. Petříček, P. Kacálek, R. Smolka, AMS **1041**, 71-74 (2014)
6. J. Plachý, *Bitumen sheet joints under the influence of temperature and time in the composition of single-membrane roofs* (Doctoral thesis, Brno University of Technology, Faculty of Civil Engineering, Brno, 2005)
7. *ETAG 006. European Technical Approval Guideline – Systems of Mechanically Fastened Flexible Roof Waterproofing Membranes* (European Organisation for Technical Assessment - EOTA, Brussels, 2000)
8. T. Petříček, *Adhesion of joints of mechanically fastened bitumen sheets* (Doctoral thesis, Brno University of Technology, Faculty of Civil Engineering, Brno, 2013)

 DOI: 10.1051/matecconf/20179303008

The impact of the year-on-year variation in the intensity of solar radiation on the energy intensity of low-energy and passive houses

Roman Šubrt[1,*] and *Pavlína* Charvátová[2]

[1]Energy Consulting, z.s. Alešova 21, 370 01 České Budějovice, Czech Republic
[2] Institute of Technology and Business in České Budějovice, Department of Civil Engineering, 370 01, České Budějovice, Czech Republic

Abstract. Solar radiation is a significant segment of heat gains in the operation of buildings. The importance of this segment is highlighted by lowering the energy performance of buildings. The current condition of assessment considers the standard values of solar radiation but these are often very different from the fair values. In the contribution it draws attention to not only to on-year variation in solar fluctuations in the intensity of solar radiation and its significant long-term deviation from the standard values but also to the impact to energy building in reliance to its energy intensity. The attention will be focused also to different values in standards valid in the Czech Republic. This specification of energy assessment of buildings is not only necessary to approximate calculations of real state, but mainly because we can expect more disputes about if a building has declared calculating the parameters of a building with nearly zero-energy or passive house.

1 Current condition in the Czech Republic

Energy performance of buildings is assessed by calculations according to valid European standards taking into consideration the climatic data in the Czech Republic. Those are given by CSN 73 0540 (CSN = Czech Technical Standard) and TNI 73 0331 (TNI = Technical Standards Information) and they should be based on long-term average. Even though TSI is just a recommended information usually the values listed there are used for assessment of buildings. Although the information differs in both standards and mainly in global radiation perpendicular surface oriented to different cardinal points. Comparison for the winter season is shown in Table 1 where there is a noticeable difference both in individual months, and within different orientation to cardinal points. These different values lead to according TNI it is less necessary to consider the orientation of windows to different cardinal points than when evaluation by CSN. Another source of values of global solar radiation for the region of the Czech Republic can be obtained from the Internet, namely from the computational programs for calculation of electricity from photovoltaic power plants.

* Corresponding author: roman@e-c.cz

In Table 2 shows the comparison of results of several houses and different assessment according to these standards.

2 Real climate data

To determine the actual status of global radiation the data were purchased from ČHMÚ (Czech Hydrometeorological Institute) which performs long-term measurements. Values for Hradec Králové are listed in Table 3 [1]. This table shows that global solar radiation over the past 10 years is higher than the long-term average which is given in CSN, although the global radiation does not correspond with average temperatures for that period.

Table 1. Comparison of values listen in TNI 73 0331 and CSN 73 0540-3.

source	Monthly dose irradiation H [kWh/(m²•month)]							
	January	Fabruary	March	April	October	November	December	**Total per year**
Horizon								
ČSN	23,06	36,75	76,12	110,53	52,74	25,53	18,62	**343,35**
TNI	20,8	37,0	72,2	113,8	56,5	25,2	14,9	**340,4**
difference	90 %	101 %	95 %	103 %	107 %	99 %	80 %	**99 %**
South								
ČSN	41,94	53,31	89,73	88,42	71,57	41,07	30,95	**416,99**
TNI	34,2	51,1	74,4	84,7	74,4	45,4	29,0	**393,2**
difference	82 %	96 %	83 %	96 %	104 %	111 %	94 %	**94 %**
Southeast, Southwest								
ČSN	32,2	41,17	76,16	84,33	57,61	31,99	23,86	**347,32**
TNI	26,8	41,0	64,7	86,4	60,3	33,8	23,1	**336,1**
difference	83 %	100 %	85 %	102 %	105 %	106 %	97 %	**97 %**
East, West								
ČSN	15,01	22,21	48,89	65,84	32,23	15,87	11,18	**211,23**
TNI	14,1	25,5	46,9	74,2	38,7	18,0	11,2	**228,6**
difference	94 %	115 %	96 %	113 %	120 %	113 %	100 %	**108 %**
Northeast, Northwest								
ČSN	6,42	9,55	23,25	38,3	14,06	6,98	5,09	**103,65**
TNI	8,2	14,8	29,8	50,4	21,6	9,4	6,0	**140,2**
difference	128 %	155 %	128 %	132 %	154 %	135 %	118 %	**135 %**
North								
ČSN	5,21	7,26	15,6	24,04	10,36	5,52	4,03	**72,02**
TNI	8,2	13,4	25,3	36,0	18,6	9,4	6,0	**116,9**
difference	157 %	185 %	162 %	150 %	180 %	170 %	149 %	**162 %**

Table 2. Comparison of assessment Building Energy Performance for heating depending on weather data according to CSN 73 0540 and according to TNI 73 0331 [GJ].

	A4	**B4**	**C4**
ČSN	32,5	31,4	30,4
TNI	31,3	30,0	29,2
difference %	4	4	4

3 Comparison of the calculated values

To determine the difference of the actual climate data was chosen reference building with dimensions of 10 x 10 m and 6 m in height with windows of 20 m^2 facing south, 4 m^2 north and after 8 m^2 on the east and west. This building was considered a total of 12 variants, according to storage capacity of construction:

A) Lightweight construction, heat capacity 110 kJ/(K·m^2),

B) Medium-heavy construction, heat capacity 165 kJ/(K·m^2),

C) Heavy construction, heat capacity 260 kJ/(K·m^2).

And also according to quality of thermal insulations. There were chosen values required by standard of 1964 [2], by standard of 1977 and also recommended values for passive houses by standard from 2011. These values are listen in Table 4, where ale also listed characteristic values of the glazing. Ventilation in the first two cases is considered natural, in the latter two cases is considered as a forced recovery.

Table 3. Global solar radiation for years 2005 - 2014 measured in Hradec Králové.

Global solar radiation E_m [kWh/(m^2•month)]									How much more sunlight strikes against CSN [%]	How many degree-days were warmer comparing to the long-term average [%]
year	01	02	03	04	10	11	12	**Total year**		
2005	28,20	43,80	100,22	135,25	73,92	22,65	15,27	**419,30**	122 %	107 %
2006	32,48	44,55	79,49	120,45	68,51	25,45	19,64	**390,56**	114 %	113 %
2007	23,99	39,45	81,48	165,44	58,55	26,35	15,69	**410,95**	120 %	126 %
2008	23,93	47,34	82,08	114,97	56,65	26,24	19,49	**370,69**	108 %	117 %
2009	24,73	33,47	64,51	162,50	41,40	29,44	17,31	**373,35**	109 %	118 %
2010	26,29	40,49	86,35	140,36	67,89	24,64	17,71	**403,72**	118 %	98 %
2011	25,21	47,80	102,62	127,84	61,22	30,11	17,63	**412,42**	120 %	118 %
2012	26,60	48,24	99,32	124,66	53,06	25,02	20,69	**397,59**	116 %	117 %
2013	18,46	34,40	76,87	105,94	67,91	24,25	20,41	**348,24**	101 %	111 %
2014	23,82	48,46	96,43	126,57	56,90	21,34	15,86	**389,38**	113 %	138 %
CSN	**23,06**	**36,75**	**76,12**	**110,53**	**52,74**	**25,53**	**18,62**	**343,35**	**100 %**	

Table 4. Considered in the calculation of thermal properties of the reference building.

construction	heat transfer coefficient U_N [W/(m²·K)]			
	CSN 73 0540 of 1964	CSN 73 0540 of 1977	CSN 73 0540-2 recommended values	CSN 73 0540-2 recommended values for passive houses
	1	**2**	**3**	**4**
on the ground floor	1,5	1,15	0,3	0,15
wall	1,4	0,89	0,25	0,12
window	2,4	2,4	1,1	0,9
flat floor	0,93	0,51	0,16	0,1
solar transmittance glass g [-]:	0,81	0,81	0,52	0,6

Table 5. Calculated energy performance of buildings according to climate conditions CSN 73 0540 and according to climate data in each years for region Hradec Králové [GJ].

	A1	**A2**	**A3**	**A4**	**B1**	**B2**	**B3**	**B4**	**C1**	**C2**	**C3**	**C4**
2005	184,7	126,1	46,9	28,8	182,2	123,5	45,1	27,7	179,2	120,2	43,2	26,9
2006	183,8	125,8	46,8	28,9	181,4	123,3	45,1	27,8	178,2	120,1	43,3	26,9
2007	164,6	111,9	40,2	25,6	162,1	109,1	38,3	24,1	159,2	106,4	37,2	23,2
2008	164,3	111,5	39,6	24,6	162,0	109,1	38,1	23,2	159,2	106,1	36,8	21,9
2009	179,5	123,1	47,0	30,0	177,3	121,0	45,5	29,2	174,6	118,7	43,3	28,5
2010	195,6	134,6	50,9	32,8	193,4	132,3	49,2	30,9	190,3	129,6	47,5	29,7
2011	178,7	121,4	44,3	26,4	176,2	118,4	42,4	25,1	173,2	115,6	40,6	23,7
2012	181,5	123,5	45,3	26,3	179,2	120,8	43,5	24,9	176,5	118,3	41,9	23,6
2013	189,1	130,2	49,0	31,1	186,6	127,7	47,4	29,6	184,4	125,5	45,9	28,3
2014	151,0	101,3	36,5	22,4	148,6	99,0	34,6	20,4	145,7	96,3	33,0	19,1
CSN	182,7	127,1	49,1	32,5	180,8	125,2	47,8	31,4	178,7	122,7	46,7	30,4

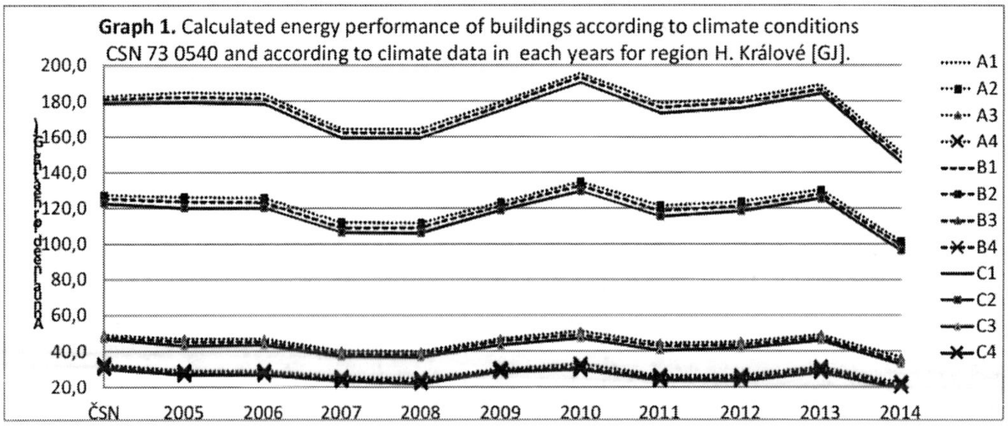

Fig. 1. Calculated energy performance of buildings according to climate conditions CSN 73 0540 and according to climate data in each years for region Hradec Králové [GJ].

DOI: 10.1051/matecconf/20179303008

Calculations of energy performance of buildings for heating were done according to CSN EN ISO 13790 and the result is 12 variants of calculation (from A1 to C4) for years 2005 – 2014. These results are listed in Table 5 and later in graph 1 [3].

In Table 6 and graph 2 are then calculated values in percentage relative to the values calculated according to the standard.

Table 6. Energy performance of buildings according to climate data in each years for the region Hradec Králové relative in percentage to values obtained under standardized climatic conditions [%].

	A1	A2	A3	A4	B1	B2	B3	B4	C1	C2	C3	C4
2005	101,1	99,2	95,6	88,6	100,8	98,6	94,2	88,3	100,3	97,9	92,5	88,4
2006	100,6	98,9	95,5	88,9	100,3	98,5	94,3	88,6	99,8	97,8	100,2	95,1
2007	90,1	88,0	82,0	78,8	89,7	87,1	80,1	76,9	89,1	86,7	79,6	76,4
2008	90,0	87,7	80,7	75,8	89,6	87,1	79,7	73,9	89,1	86,5	78,6	72,2
2009	98,3	96,8	95,8	92,3	98,0	96,7	95,2	92,9	97,7	96,7	92,6	93,9
2010	107,1	105,9	103,8	101,1	106,9	105,7	102,8	98,3	106,5	105,6	101,6	97,7
2011	97,8	95,5	90,4	81,4	97,4	94,5	88,7	80,0	97,0	94,1	86,8	78,0
2012	99,4	97,1	92,4	81,1	99,1	96,5	91,0	79,4	98,8	96,4	89,6	77,6
2013	103,5	102,4	99,9	95,7	103,2	102,0	99,1	94,3	103,2	102,2	98,2	93,2
2014	82,7	79,7	74,5	68,8	82,2	79,1	72,3	65,0	81,5	78,5	70,6	62,9
CSN	100,0	100,0	100,0	100,0	100,0	100,0	100,0	100,0	100,0	100,0	100,0	100,0

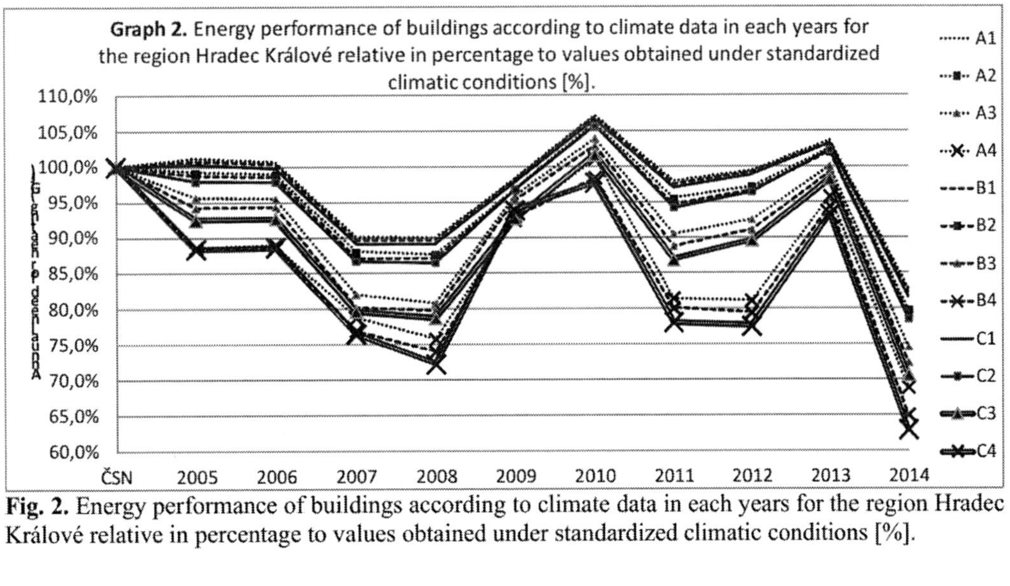

Fig. 2. Energy performance of buildings according to climate data in each years for the region Hradec Králové relative in percentage to values obtained under standardized climatic conditions [%].

Effect of changes in the intensity of solar radiation on energy consumption for heating is documented in Table 7 and graph 3. There is calculated energy consumption for heating according to values listed in the standard and then is considered the same temperature, however the intensity of solar radiation is higher by 20%.

Table 7. Energy performance of buildings according to climate data according to the standard and increase in the global solar radiation by 20 % for the region Hradec Králové listed in GJ and in % relative to standard values.

	A1	A2	A3	A4	B1	B2	B3	B4	C1	C2	C3	C4
CSN [GJ]	182,7	127,1	49,1	32,5	180,8	125,2	47,8	31,4	178,7	122,7	46,7	30,4
Temperatures according to CSN, sun more by 20% [GJ]	176,6	121,1	45,7	29,3	174,4	118,2	44,3	28,1	171,8	116,1	42,8	27,0
Likewise %	103,5	105,0	107,3	111,0	103,7	105,9	108,1	111,9	104,0	105,8	109,3	112,7

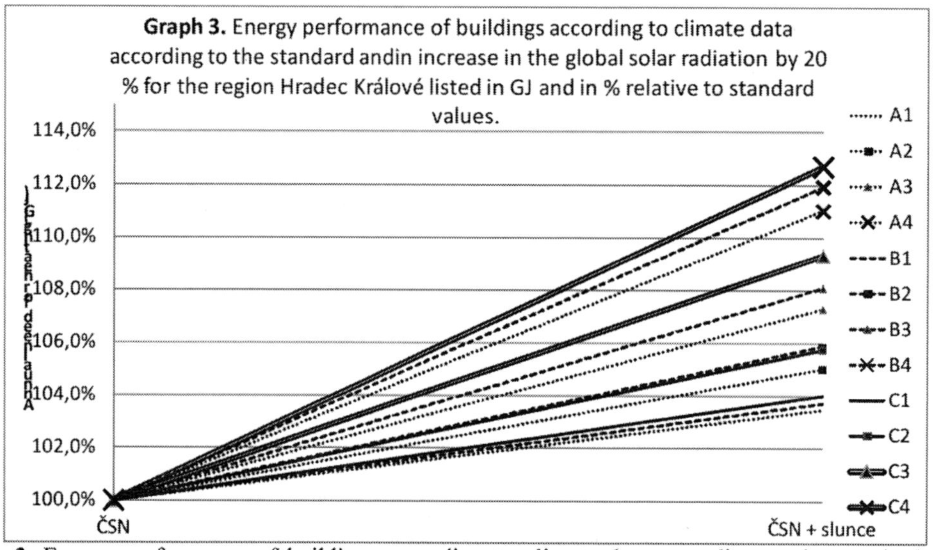

Fig. 3. Energy performance of buildings according to climate data according to the standard and increase in the global solar radiation by 20 % for the region Hradec Králové listed in GJ and in % relative to standard values.

4 Conclusions

The results demonstrate clearly that the effect of fluctuations in solar radiation has a significant influence to energy performance of buildings for heating which increases especially with thermal insulation properties structure. In old buildings, the 20 % difference in the global radiation shows at model building the range of 3,5 to 4% depending on the heat capacity of the considered structure. For buildings with thermal insulation as recommended for passive houses, however, the difference is 11 to 12,7%. From this finding it implies knowledge that when assesing the measured energy consumption for heating is necessary, besides other things, také in account also global solar radiation, which is usually not performed, usually only takes into account the different temperatures in the heating season. Less significant impact on the energy performance of buildings have storage capabilites. Those at the model house ranged to 2,3%.

References

1. R. Šubrt, P. Charvátová, Climatic data from 07/2014 to 06/2015 (AES, 2015)
2. *ČSN 73 0540, Thermal protection of buildings* (1964)
3. R. Šubrt, P. Charvátová, *Katalog tepelných vazeb II – vnější stěny* (Isover, Prague, 2015)

DOI: 10.1051/matecconf/20179303009

Coordination of public infrastructure solution for preventing of accidents

Václav Jetel[1,*]

[1]Czech Technical University in Prague, Faculty of Civil Engineering, Department of Urban Design, Town and Regional Planning, Thákurova 7, 166 29 Praha 6

Abstract. Coordination on public infrastructure is an integral part of any town and country planning process, which aims to provide unlimited uptime of all systems, providing basic needs in settlements that cannot cause outages, failures and accidents of the systems themselves, threaten public safety or cause defects or even destruction of the surrounding buildings. The paper describes the selected policy coordination on technical equipment area, which monitors the stability of the surrounding buildings, eliminating the potential defects, using scientific and educational knowledge in the field of municipal engineering. In the end, against the backdrop of actual accidents and failures of utilities, can be documented need for a comprehensive approach to the design of public infrastructure, already in the planning preparation in approving land-use plans, regulatory plans and decision-making in the planning stages.

1 Introduction

Coordination on public infrastructure is one of the most important activities of their operators, municipal engineers, architects, urban planners, economists, environmentalists, politicians and concerned government authorities. That this activity is properly implemented, it must be set within a framework of rules laws and their implementing regulations, standards, technical requirements and guidance, based on claims on public infrastructure. One of the missions of the Department of Urban Design, Town and Regional Planning, in Czech Technical University in Prague, then prepare professionals for the building industry, who will be able in a variety of jobs to seek optimal solution for the future construction of public infrastructure so as not to limit its operability, and while that may not cause outages, failures and crash the systems themselves, threaten public safety or cause defects or even destruction of the surrounding buildings. Already in a planning process, which this article focuses, can significantly affect the future behaviour of buildings and structures for eliminate any malfunctions. An important instrument for coordinating public infrastructure solutions have become the planning analytical materials, whose main theme is just preventing threats in the area, which could cause the collapse of individual buildings.

* Corresponding author: vaclav.jetel@fsv.cvut.cz

2 Public infrastructure

Infrastructure is the basic facilities, services, and installations needed for the functioning of a community or society, such as transportation and communications systems, water and power lines, and public institutions including schools, post offices, and prisons. The design and engineering of infrastructure was historically conceived in isolation, independent of the overall urban vision. Infrastructural system is often hidden from view, apparent, which breeds unhappy surprises and mounting frustration when the system fails. [1]

In the frame of legislation Planning and Building Code, public infrastructure are land, buildings and transportation infrastructure facilities, technical infrastructure, civic amenities and public spaces, and to set up and used in the public interest.

Table 1. Public infrastructure under the Building act.

Transport infrastructure	for example, the structures of roads, railways, waterways, airports and the facilities related to them [2]
Technical infrastructure	the lines and the structures and to them operationally related facilities of technical equipment, for example, water mains, distribution reservoirs, waste water sewage, waste water treatment plants, structures and facilities for waste disposal, transformer stations, power distribution lines, communication lines of public communication network and electronic communication equipment, product pipe lines [2]
Public services	the structures, facilities and lands serving, for example, for training and education, social services and care of family, health services, culture, public administration, protection of inhabitants [2]
Public space	all squares, streets, marketplaces, sidewalks, public greenery, parks and other areas accessible to everyone without restriction, ie for general use, regardless of ownership of the space [3]

3 Coordination of technical equipment area

Activities under the coordination of the technical facilities of the area can be defined as the ability to file joint solutions to problems, problems of various kinds of utilities, interrelated, which is necessary to deal preventively to avoid especially for our environment.
Coordination can be divided into several following topics:
- spatial,
- technical and economic,
- technology,
- technical and ecological,
- coordination on their mutual direct functional links or links to other elements,
- coordination ways of addressing modes and load, flow systems, reliability, security, technical documentation solutions and more.

In planning activities apply to individual coordination of spatial planning instruments, which are defined in the Building Act and which have different implications in the context of specific outputs.

Each instrument has its own territorial scope, scale and detail solutions to problems ranging from the most general nationwide (Spatial development policy) to detail the placement of buildings (regulatory plan).

4 Selected town and country planning instruments

4.1 Planning analytical materials

The planning analytical materials contain the ascertainment and assessment of the state and development of the area, its values, limitation of the changes in the area due to protection of public priorities, arising from the regulations or stipulated under the special regulations or arising from the properties of the area, programmes for executing the changes in the area, ascertaining and assessing the area sustainable development, and determination of problems for solution in the planning documentation [2].

To coordinate the territories currently have a pivotal role, since they give into the hands of designers and procurers of the necessary elements that define when a thorough analysis of possible problems in the concept of public infrastructure and set tasks for future planning activities. For the purpose of eliminating threats are the most important of these data on the area: undermining areas, landslide areas, contaminated areas, areas with poor air condition, floodplains, emergency zones, security zones and noise zones. In this year, taking place for the third full update planning analytical materials to evaluate current trends of development of the territory and its society since 2008.

4.2 Planning study

Planning study suggests, examines and considers possible solutions of selected problems, or arrangements or the development of certain functional systems within the area, for example, the public infrastructure, territorial system of ecological stability, which could significantly impact or condition the utilization and arrangement of the areas or of their selected parts [2]. Currently, the main topics are studies of landscape and studies of public space. Now a study of public spaces may significantly affect the future conduct of public infrastructure in settlements, solve their problems and eliminate the threat.

4.3 Spatial development policy of the Czech Republic

The spatial development policy (SDP), with respect to the possibilities of the area, coordinates creation and updating of the development principles, creation of concepts approved by ministries and other central administrative authorities, and programmes for the changes in the territory of the national wide importance, and determines the tasks ensuring this coordination [2]. SDP coordinates also intentions to make changes to the area for transport and technical infrastructure that impacts territories of several regions due to its significance, extension or foreseen usage. SDP delimit areas with development potential within inundation spaces and locate public infrastructure in there, fi nanced from public budgets, in exceptional and well reasoned cases only. Delimit and protect areas with development potential for relocation of structures from spaces with high flood damages risks.

4.4 Development principles

Development principles within the supra local context of the region´s territory specify and develop the objectives and tasks of the town and country planning in accordance with the spatial development policy, they determine the strategy for their fulfilling and they coordinate the planning activities of municipalities [2]. The importance of coordination of development principles for example in Prague completely suppressed and the Prague Institute of Planning and Development rather understands the importance of setting priorities for city development and definition of the construction of transport and technical infrastructure are important for the function of the city as a whole.

4.5 Master plan

The master plan determines the basic concept of the development of the municipality, protection of its values, its areal and spatial arrangement, arrangement of the landscape, and the concept of the public infrastructure; delimits the developed area, areas and corridors, especially the areas with development potential and the areas delimited for the alteration of the existing development, for redevelopment or repeated utilization of the depreciated area, for public works, for public benefit measures, and for the territorial reserves and determines the conditions for utilization of these areas and corridors [2]. Concept of public infrastructure is one of the most important concepts in the master plan, which already includes specific solutions to problems that can cause a real threat in the area. Measures to eliminate threats in the public interest and can restrict the right to property. An important part of the master plan is defined territorial measures in case of an emergency or natural disaster. Often, however, these measures zoning plan completed only formally.

4.6 Regulatory plan

The regulatory plan within the settled area determines the detailed conditions for the use of the grounds, for location and spatial arrangement of structures, for protection of values and character of the area, and for creation of a favourable environment. The regulatory plan always determines the conditions for delimitation and the use of the grounds, for location and spatial arrangement of the structures of the public infrastructure, and delimits the public works or the public benefit measures [2]. Thanks to detail with which the regulatory plans are working (at the level of planning permission), can specifically identify and locate such measures that eliminate potential threats in specific locations. Here we use the services of specific standards, which set rules for the placement of public infrastructure works, which are defined so as to eliminate potential threats.

5 Conclusion

Infrastructural systems operate in the background, people know that they exist, but would rather forget about that. We are reminded of their existence when health and safety are at risk [1].

At the time this article was being prepared (26. 7. 2016), took place in Prague, Vysocanska Street, to drop the road where inside this crater of about 8 x 4 m rammed passenger car. The road, by the Prague Water and Sewerage, collapsed after torrential rain due to the crash sewer. Water hollowed out cavern under the roadway, which caused the collapse of several meters of brick sewer ovoid shape. Damage to property was estimated at tens of millions of crowns. Cavern managed to stabilize the backfill, compacted and conclude concrete slab, which then allows repair brick sewer. This event illustrates the power of water, which can cause destruction of buildings and equipment unless they include in the draft utilities eventual elimination of possible threats, in case of floods, accidents, poor sewerage system, including a proposal for its implementation.

Accidents utilities should be eliminated in its occurrence appropriate precautions to avoid loss of human lives, environmental damage and economic losses.

One of the 10 largest industrial disasters after the 2nd World War was for example the gas pipeline accident in Guadalajara, Mexico, which caused gas leaks into the sewer system and subsequent explosion, the result of which was destroyed entire neighbourhoods. It was thus a combined disaster two utilities, which had the character of the use of weapons of mass destruction.

Table 2. Top industrial disasters (1945-1990) [4].

Year	Location	Type	Deaths
1984	Bhopal, India	Toxic vapour / methyl isocyanate	2750-3849
1982	Salang Pass, Afganistan	Toxic vapour / carbon monoxide	1500-2700
1956	Cali, Colombia	Explosion / ammunitions	1200
1958	Kyshtym, Russia	Radioactive leak	1118
1947	Texas City, USA	Explosion / ammonium nitrate	576
1989	Acha Ufa, Russia	Explosion / natural gas	500-575
1984	Cubatao, Brazil	Explosion / gasoline	508
1984	St Juan Ixhautepec, Mexico	Explosion / natural gas	478-503
1992	Zonguldak, Turkey	Mine explosion / gas	388
1983	Nile River, Egypt	Explosion / natural gas	317
1992	Guadalajara, Mexico	Sewer explosion / gas	210
1986	Chernobyl, Ukraine	Explosion / radioctivity	31-300

On the basis of these disasters and the large number of accidents with impacts to the environment we must still ask whether our public infrastructure administrators are able to ensure future safe operation of infrastructure that cannot be repeated similar events and that have all the systems under control. For example, most accidents in public infrastructure are associated with energy production and distribution, which is a tax on economic growth [5].

References

1. *Landscape infrastructure: case studies by SWA* (Birkhäuser, Basel, 2011)
2. *Act no. 183/2006 Coll., Building Act* (2006)
3. *Act no. 128/2001 Coll., Municipality Act* (2011)
4. S.L. Cutter, *Hazards Vulnerability and Environmental Justice* (Routledge, 2012)
5. P. Šrytr, *Výpadky inženýrských sítí z hlediska udržitelného rozvoje* (Czech Technical University in Prague, Prague, 2009)

Categorization of constructional and technical condition for processing of brownfield identifying studies

Zuzana Kramářová[1,*]

[1]Faculty of Technology, The Institute of Technology and Business, Okružní 517/10, 370 01 České Budějovice, Czech Republic

Abstract. The issue of brownfield revitalization is one of very actual topics in the area of the sustainable development and construction. The registration of localities characterized as brownfields is addressed within compiling and updating Land Analytic Background of municipalities in the Czech Republic. However, it is very unsatisfactory for practical revitalization needs, and it is partly due to disunity in the data recorded in different locations in different administrative territories, and also the small range of such data. Therefore, so called "Brownfield Identifying Studies" are important to compile in particular administrative districts at the level of municipalities with extended jurisdiction. These studies are tasked with not only identifying brownfields = localizing but also keeping database of them with closer data which can make potential investor's decision much easier. Constructional and technical condition is one of the assessed and registered data when compiling and cataloguing Brownfield Identifying Studies. As these databases are used by large - unprofessional public it is necessary to describe the condition as simply as possible, for instance to express it by a numerical marking. The pitfalls of brownfield constructional and technical condition classification are described in detail in the contribution.

1 Introduction of Brownfield Issue

The issue of brownfield revitalization is one of the very actual topics in the area of sustainable development and construction. The first step of successful revitalization is timely compiling all available information about a certain location, eventually chosen locations in order to be possible to eliminate negative effects of the certain location in time, e.g. area contamination, bad constructional and technical condition of the objects, inappropriateness of the new functional usage with respect to its surroundings, …For these purposes it is useful to register brownfields in the area and when the conditions of one of the brownfield are not convenient, another object can be brought in to a potential investor who can be protected from leaving the certain area.

The registration of localities characterized as brownfields is addressed within compiling

* Corresponding author: 19511@mail.vstecb.cz

and updating Land Analytic Background (UAP) of municipalities in the Czech Republic. This is legally enshrined in Law no. *183/2006 Sb on Planning and Building Code (Building Law)* [1], especially in its processing *Decree No. 500/2006 Sb., on Land Analytic Background, Urban Planning Documentation and Register Methods of Planning Activities* [2] where there are observed phenomena listed and updated in UAP of municipalities in annex 1 part A. Brownfields belong among phenomena no.4, "areas to recover or reuse of devaluated territory". UAP update takes place every 2 years in the Czech Republic as appears from Building Law.

However, it is very unsatisfactory for practical revitalization needs, and it is partly due to disunity in the data recorded in different locations in different administrative territories (a processor differs from a processor, even Administrative District differs from another Administrative District – according to available finance spent on UAP update), and also the small range of such data. This range is usually limited to a name of location, serial number, cadastral territory and plot number, original use and constructional and technical condition of the location. In many cases there are not known even all these most basic data. Therefore so called "Brownfield Identifying Studies" are important to compile in particular administrative districts at the level of municipalities with extended jurisdiction. These studies are tasked with not only identifying and localizing brownfields (which means to discover them) but also keeping database of them with closer data which can make potential investor's decision much easier [3].

2 Brownfield Identifying Studies

Study for Locating Brownfields [4] was one of few studies done across the board. It was co-financed by Business and Investments Support Agency CzechInvest. The study was processed on the whole territory of the Czech Republic in years 2005 – 2007. However, not only the different level of processing in different regions was its main insufficiency because the particular regions were except CzecInvest the project co-owners that did not unify observed information about particular locations but also its very most fundamental commission which was: "Find 200 brownfields in the region which floor area is larger than 500 m^2 or ground area is larger than 1ha ". From this commission it is clear that the demand was not to find all brownfields meeting the asked parameters but exactly 200. In some regions this led to integrating of properties that could not be definitely considered brownfields, somewhere else it happened that the number of locations was reduced (they were eliminated afterwards) or the research ended when the defined number was reached (a part of the territory was not been explored). Nevertheless, basic mapping of the issue condition on the whole territory of the Czech Republic was the indisputable benefit of the study including collection of more detailed information about the individual locations. In some regions this collection took place in several phases, in other ones conversely one phase only was done. For instance, in Region Hradec Králové there were 3 phases set in which previously got information was specified and searched again in every further phase. South Bohemian Region was an example of one phase only data collection there was collected more detailed information in one step but less amount only there. The study outputs were one of the sources at processing of National Brownfield Regeneration Strategy. Unfortunately, it has not been approved by the Government of the Czech Republic yet.

3 Categorization of Constructional and Technical Condition of Brownfields

Basic purpose of the identifying studies is to gain a compact brownfield file = reusable areas which will be introduced to potential investors as an offer, and due to this file annexation of open landscape, especially agricultural land and valuable natural areas will be decreased. Nevertheless, it is not required to publish all data directly from a practical (data size) point of view and with respect to personal data protection. From experience at processing of brownfield identifying studies intended for practical needs in territory, multistage cataloguing demand has clearly arisen which means to create at least two-stage database. The first stage should be absolutely available for public, paper as well as electronic version which is remotely accessible would be ideal. This stage should contain basic information about the property/location, (lit. dissertation) – name, serial number, type, cadastral area, size/floor area, number of objects, condition primary evaluation, basic information about traffic and technical infrastructure, information about contamination, ownership type and information about greenness, and all completed with approximately 4 characteristic photographs and map position plot. The second stage should not be made open to public without an interested person being registered in advance because it can contain sensitive data as owner's phone numbers, e-mail contacts, information about liens, ... This stage has to more concretize the previous information thus for instance technical infrastructure concerning it is necessary to describe not only concrete connections existence but also their eventual capacity or nominal diameter, distance to possible connecting or connection lengths and their trace, eventual owners (when they are different from the brownfield owners), ...

Table 1. Brownfield constructional and technical condition classification scale at first cataloguing stage.

evaluation	verbal evaluation	detailed description
1	excellent condition	brownfield is possible to use immediately without any change of functioning and any invests in repairs
2	good condition	brownfield needs minor building modifications or common maintenance work that can be postponed eventually (e.g. remediation of facade)
3	worse condition	most common condition – brownfield needs building repairs (e.g. water spout exchange, inner water and drain piping repair, electrical wiring, ...)
4	bad condition	brownfield cannot be used without large repairs (e.g. roof timbers repair, remediation of damp supporting structures, windowpanes, door panels, floors and interior wall plasters, ...
5	critical condition	brownfield is in serious disrepair – supporting structures are structurally unsound, making large complex repair or demolition should be considered

Description of brownfield constructional and technical condition is one of important area of observed data especially for investors. Its processing should be integrated into both stages of catalogue information. Working on the assumption that the first stage database is used by large unprofessional public, it is necessary to describe such condition at first stage as simply as possible, for instance to express it by numerical marking. Most commonly used classification scale is marking form 1 to 5, see Table 1.

The disadvantage of this scale is that almost 2/3 of brownfields get mark 3 – worse condition. Though the range of this category is large – from objects in relatively good condition that need most necessary repairs only for instance as interior wiring and piping up to objects quite devastated that need immediate intervention because atmospheric precipitations are leaking inside and that leads to initial damage of interior equipment, wiring, piping and floors. Other relatively common grade is grade 5 - critical condition. The other conditions are represented rarely.

Table 2. Example of evaluation of object condition.

object parts			evaluation	description/note
roof	covering and hydro-isolation		2	covering changed in 2001
	supporting construction		1	timbers reconstruction in 2001
	rain water drainage (spouts + sheet metal covering)		2	spouts OK, need to clean
	outlets(chimney heads, breather chimneys, ...)		3	need to re-point and remediate chimney slab
upper constru ction	main supporting structures	horizontal	2	beam ceilings – need to renew paints against pests
		stairway	2	wooden – need to renew paints against pests
		vertical interior	1	brick
		perimeter	1	brick
	perimeter coat		-	-
	plasters	outside	4	north side façade is falling away, the others have evident bulges
		interior	3	ground floor is damaged by graffiti artists, the other floors need to repair partially – rare bulges.
	floors		3	wooden – on ground floor significantly damaged – need to change, on the others can be repaired
	panes and panels	window	4	significantly damaged on ground floor by vandals, the others can be repaired by glassing
		door	5	need to change, significantly damaged
		others	-	-
	technical building equipment installation	water piping	5	cracked by frost
		sewerage	3	partially in order, branch with vertical drainpipes blocked
		electro	5	need to install all new – not fitting current demands
		heating	4	central Heating – cracked by frost

		gas	-	-
		others	-	-
	security elements (banisters, …)		2	forged – need to clean surface rust but firm
	chimney units		2	solid fuel – need to clean
	balconies and other constructions		-	-
	other equipment and components		-	-
lower construction	underground floor		-	none
	lower construction hydro isolation		3	outdated and ineffective
	basement construction		-	not identified

4 Conclusion

Concluding it is possible to summarize that Brownfield Identifying Studies if they are processed for administrative areas, they are an important instrument of land protection against open landscape annexation. When they are processed properly they can serve not only as background for Land Analytic Background (UAP) processing but also as an offer for potential investors. Evaluation of constructional and technical condition of locations as whole and also more detailed classification of their individual parts is one of the most important part of these studies. This evaluation definitely affects pre-projection decision of investors and its quality can significantly affect invest costs amount – expected as well as unexpected.

References

1. *Law no. 183/2006 Sb on Planning and Building Code* (Prague, 2006)
2. *Decree No. 500/2006 Sb., on Land Analytic Background, Urban Planning Documentation and Register Methods of Planning Activities* (Prague, 2006)
3. Z. Kramářová, *Ecological, functional, urban and architectural problems of brownfields in small towns* (Doctoral thesis, 2006)
4. Atelier T-plan, et al., *Vyhledávací studie pro lokalizaci brownfields na území Královéhradeckého kraje* (The Main Project, 2006)

AUTHOR INDEX